Information Theory

Information Theory

A Guide in Understanding Health and Disease

James Morris

Contents

Chapter III

Infection

Chapter IV

Ageing

Chapter V

Health and Disease

Chapter VI

Sexual Attraction

Chapter VII

Overview

Preface

This is a book about ideas; ideas based on fact but transcending fact. It is an attempt to provide a theoretical framework for understanding health and disease and an attempt to pose questions to guide health research. We are all interested in preserving our own health and that of our families and in preventing disease; this applies just as much to the person who places a pound coin in the charity box as it does to a top international scientist leading research into cancer or heart disease. This book is for both of them and all those in between.

Rutherford said that there are two branches of science; there's physics and there's stamp collecting.

In physics, theory guides experiment. Theoretical physicists develop theories based on ideas and use mathematics to argue logically from the idea to a testable question or a testable prediction. Experimental physicists then test the idea against reality in a carefully designed experiment. If the theory accords with reality it can be developed further; if reality contradicts the theory then in needs to be modified and then subject to further tests. Those who are good at theory, and at mathematics, are not necessarily good at experiments. Those who are good at designing and performing experiments are not necessarily good at theory. But together they succeed.

In biology and in medicine we tend collect data and then ask "what does it mean"? Any theory is *post hoc*. Theory does not guide research but follows research. This is what Rutherford meant by stamp collecting.

It is my contention that just as there is a theoretical physics then there should be a theoretical biology. Just as there are theoretical physicists then there should be theoretical biologists who develop theories using the intellectual rigour of mathematics and put forward those theories for others to test. The title of the book indicates that this theoretical biology should be based on information theory. It is suggested that information is to biology what energy is to physics; a fundamental idea on which theory can be constructed.

Preface

"What is energy"? It is a fundamental concept; it has no component parts and therefore we cannot define it, but we can measure it. It is a scientific concept because it can be measured. Much of physics is based on theories developed using the concept of energy. "What is information"? It is a fundamental concept with no component parts so we cannot define it, but it can be measured. Information is measured as the reduction of uncertainty. One bit of information reduces uncertainty by one half. Information and uncertainty are closely related ideas; they are two sides of the same coin.

When information is processed the following rules apply:

- Information processing systems have a finite capacity.
- There is a finite chance of error.
- Systems decay at random according to the laws of entropy and therefore the probability of error rises with time.
- All complex information processing systems have a high level of redundancy in order to reduce the error rate.

The laws are explained in the first chapter and developed in subsequent chapters.

The way we age matches the way in which a complex highly redundant system subject to the laws of entropy deteriorates with time. As we grow old the probability of disease and death is initially low but then accelerates rapidly towards the end of our lifespan. This is exactly as predicted if we are regarded as a complex highly redundant system subject to random insults from the environment. Death, disease and impairment of health can be explained in terms of an interaction between a potentially threatening environment and a complex system attempting to avoid environmental insults. The system is successful most of the time but is not perfect; errors will occur and the probability of error will rise with time. Redundancy reduces the error rate so that the chance of a mistake is low in youth and middle age but rises rapidly in old age.

In the developed world the frequency of asthma, eczema, hay fever, type 1 diabetes mellitus, multiple sclerosis, Crohn's disease, rheumatoid arthritis and other disorders of immunity are on the increase. The favourite theory is the "hygiene hypothesis"; that in some way changed exposure to germs because of better hygiene is increasing the risk of a disordered immune response. This hypothesis is vague and *post hoc* but sadly typical of theory in medicine. In chapter 3 I develop the hygiene hypothesis from first principles using information theory. This more precise hypothesis leads to the prediction that type 1 diabetes mellitus is caused by delayed exposure to common bacteria usually resident in the gastro-intestinal tract. This is contrary to the accepted view that if infection is involved it is epidemic viral infection rather than commensal bacterial infection. Time will tell; but this demonstrates the power of theory to make precise testable predictions.

The ideas on hygiene and immunity are derived by regarding the immune system as a complex highly redundant information processing system deteriorating with time. We meet the majority of common bacteria and viruses early in life when our immune system works well. But if exposure is delayed by better hygiene the chance of a mistake on first exposure rises. One type of mistake is to fail to realize that a protein on the organism resembles one of our own body proteins; a reaction to attack the protein then leads to an attack on our own tissues (autoimmune disease).

These considerations lead to the general conclusion that we ought to strive to ensure early, low dose, mucosal exposure to common micro-organisms. In that way the chance of infection and immune disease will be reduced to a minimum. I regard this has one of the big challenges to medicine and to public health over the next 50 years.

The ultimate store of information that specifies man and woman is in our DNA. Four bases taken three at a time, (triplets), code for amino acids. A gene is a string of triplets that codes for a string of amino acids which together constitute a protein. There are 20 different amino acids and a total of approximately 25,000 different genes. An intermediate string of RNA is copied from the DNA and in turn guides the assembly of the string of amino acids. By cutting and splicing the RNA in different ways it is possible to produce over 250,000 different proteins from the 25,000 genes. Our complexity is the product of these 250,000 proteins which is known as our proteome. The proteome directly interacts with and learns from our environment. Our anatomy, physiology and metabolism are the proteome in its myriad forms.

Preface

When DNA is copied in cell division there is a finite chance of an error occurring leading to a change in the sequence of bases; this is a mutation. If a mutation occurs in a DNA sequence that codes for a protein there is a good chance that the altered protein will not work i.e. the gene has been effectively lost. This is a deleterious mutation. If the altered protein functions as well as the previous protein the mutation is neutral. If a deleterious mutation arises in spermatozoa or oocytes (the female eggs) then it can be passed to the next generation. A steady build up of deleterious mutations in the genome is prevented because the deleterious mutations interact synergistically to degrade performance of all systems (another prediction of information theory). Thus some zygotes that form from the fusion of spermatozoa and oocytes do not develop because they have too many deleterious mutations.

The zygotes that ultimately develop into adults can have any number of deleterious mutations between 0 and 16 but the average is approximately 8 (see Ch. 1). In technical terms the distribution of deleterious mutations is a slightly skewed Poisson distribution with a mean of 8. If deleterious mutations interact synergistically to degrade performance of all systems, which information theory indicates, then we are led to the most controversial prediction: that the number of deleterious mutations in the genome directly influences intelligence, health and physical attraction. Those with few deleterious mutations in the genome will, all other things being equal, be more intelligent, better looking and healthier than those at the upper end of the Poisson distribution.

Deleterious mutations and intelligence are discussed in chapter 2. The possible role of deleterious mutations in physical attraction is covered in chapter 6. Deleterious mutations and health are dealt with in the other chapters. A very positive idea to emerge from this analysis is that reduction of the mutation rate would lead over several generations to an overall fall in the number of deleterious mutations in the genome. If this occurred the entire population would be healthier, more intelligent and better looking. The way to achieve this is to work out which aspects of our environment and our lifestyle influence the mutation rate with a view to subsequent intervention; another big role for research in the next 50 years.

There is more to health than the absence of disease. People can suffer ill-health without having a diagnosable medical disease. Being overweight, underweight, unhappy, discontented, unfit and suffering symptoms classified as psychosomatic are all conditions of less than optimal good health but not disease in the usual sense of the word. Furthermore a high risk lifestyle is high risk for ill health and disease but the behaviour itself is not a disease. Thus there are important aspects of health that lie outside the usual remit of the medical profession. This area is increasingly being defined as health science and the view is growing that while bio-medical sciences underpin medicine it is the behavioural sciences that underpin health. My own view is that doctors should be concerned with the whole of health not just disease and should rely on behavioural science as much as bio-medical science. But I acknowledge that there is a growing army of non-medical health scientists who have an important contribution to make. They will find information theory just as valuable as will pathologists like me who concentrate on disease.

Alan Turing, a key figure in the development of the digital computer, was interested in the mind. He thought the mind was a machine and he wanted to produce a machine that could think. As a result one of the first groups of biologists to be interested in information theory was the psychologists. This interest, however, waned after the initial enthusiasm but it is now time for a revival. Using information theory and decision theory we are led to conclude that the brain will influence the body just as the body can influence the brain. Furthermore decision theory indicates that values and costs associated with right and wrong decisions made in an uncertain world will influence the actual decision if an optimal decision strategy is followed. In other words our perception of our own self worth will influence the multiple decisions made each day by our proteome including such actions as that of a T lymphocyte inspecting a foreign protein on the surface of a macrophage. Thus the way we feel about ourselves and view ourselves will influence the molecular response to infection. This is counterintuitive but it is a prediction of decision theory. Is this health science or is it medical science? These ideas are discussed in chapter 5 in the section on health inequality.

Medical students have to learn a vast number of facts before they can safely diagnose disease. The subjects they must master include anatomy, physiology, biochemistry, molecular biology, pharmacology, pathology and now sociology and psychology. Non-medical health scientists have a deep professional knowledge of some of these subjects but find it difficult to get an overview of the rest. The general public is interested in health and disease and they have a right to know as they fund it; but obviously they are at a great disadvantage in trying to gain a reasonable overview of the subject. This book is aimed at all these groups, medical scientist, health scientist and the general reader. In writing it I have tried to keep the number of facts to an absolute minimum and to keep terminology as simple as possible. The facts that I use in the book are all generally accepted within the medical and health sciences professions and they are quoted without references to keep the text as simple as possible. Although the facts are not contentious the ideas, hypotheses and theories that I develop from the facts are contentious. I hope it is clear in the text what is fact and what is theory.

Some will find my views on the role of genetic influences in intelligence, health and physical beauty not just contentious but inflammatory. I would ask them to consider the arguments carefully because I believe that once a clear distinction is made between neutral and deleterious mutations the potential for offence in linking genetics to intelligence disappears. The difference between races depends on neutral mutations; they are not better or worse but different. They are not the basis of any difference in intelligence as measured by the IQ test. The genetic differences within a race are determined more by deleterious mutations, they interact synergistically to impair performance of all aspects of biological behaviour. It is suggested that genetic variation in IQ scores depends on deleterious mutations.

The suggestion of a link between deleterious mutations and physical beauty is even more contentious, it is certainly counterintuitive, perhaps it is just wrong, but I doubt it. There is an enormous biological advantage in producing progeny with a low number of deleterious mutations and the best way of achieving this is to find a partner who has a low number of deleterious mutations. Deleterious mutations interact synergistically to impair all aspects of biology including the development of a symmetrical body form. Thus beauty and low levels of deleterious mutations are linked and we recognize it. I go much further in chapter 6 in developing this theme and come to the conclusion that aesthetics is much more important than utility in choosing a partner and that in nature it is the most beautiful that survive.

The research themes that emerge are:

- Measuring and monitoring rates of somatic mutation with a view to decreasing the overall rate in the population. If successful this would lead to a healthier population in future generations and would slow the rate of ageing in the current population.
- Determining the composition and kinetics of the microbial flora in much greater detail with a view to ensuring early, low dose, mucosal exposure.
- The analysis indicates that we are more interested in aesthetics than in utility, for deep biological reasons, and if maximizing happiness is an aim of health policy we need to focus on non-economic factors.

Chapter One

Information Theory

Energy and Information

The most famous equation in science is $E = mc^2$, it shows the relationship between energy (E), mass (m) and the speed of light (c). The equation is famous because it's Einstein's equation and he is the most famous scientist of recent times; his image is synonymous with scientific genius. The equation is also famous because it underlies the nuclear age and the nuclear bomb. The ability to convert matter into energy, a prediction of the equation, led to the tragedies of Hiroshima and of Nagasaki and cast a cloud of fear over the second half of the twentieth century. The third, and probably the most important, reason for the fame of the equation is that it indicates, or at least purports to indicate, that the fundamental constituent of the universe is energy. This has proved to be one of the most successful ideas in physics; based on this concept it is possible to explain the development of the universe from the moment of the big bang to the present epoch. But what exactly is energy? It cannot be defined because it is a fundamental concept and has no component parts, but it can be measured and we are familiar with the ideas of kinetic energy, potential energy, chemical energy, heat and work.

Energy is an abstract concept, but when we encounter energy the experience is real and tangible. If a stone is dislodged from above and falls on one's head, the potential energy of its elevated position is converted into the kinetic energy of motion and when it lands it hurts. Potential energy and kinetic energy are mathematical constructs but the pain is real. Equally chemical reactions and interactions can release or absorb energy; abstract ideas, but real when you consider that chemical reactions can create explosions leading to loss of life and limb. If you kick a stone the object is tangible, but the idea that it contains enormous potential energy measured by the equation $E= mc^2$ is an esoteric concept divorced from normal experience. Energy is consumed during work and released as heat; the first concept is abstract but the second and third are part of normal experience.

In the second half of the twentieth century the pre-eminence of energy has been challenged by the concept of information. Information is also fundamental and therefore cannot be defined because it has no component parts; but it can be measured. The unit of information is the bit; it is log to the base 2 of the reduction of uncertainty. Let us imagine a situation in which a crime has been committed and there are 1000 suspects, 500 male and 500 female. There is considerable uncertainty about the culprit. If new information indicates that the culprit was male rather than female then uncertainty has been reduced by one half and this is equivalent to one bit of information. Formally; Information in bits = $\log_2$ (initial number of possibilities/ final number of possibilities), assuming that all possibilities are equally likely. If the possibilities are not equally likely the formula is more complicated but not different in principle.

The central argument advanced in this book is that information is to biology what energy is to physics. We have a deeper and clearer understanding of the physical world because of theory based on energy. In this book I hope to show that information can do the same for biology.

A Brief History of Information Theory

Information has been spread by word of mouth, by letter, by book, by poem and by story since ancient times. The idea of information is not new but in the nineteenth century new modes of communication were introduced including the telegraph and the telephone. The engineers who invented these new systems of communication needed to define more precisely and to measure whatever it was they sent, often at great expense, along their wires.

In the nineteenth century the centre of Australia was explored and a route found from Adelaide in the south to Darwin in the north. The main impetus for this exploration was to construct a telegraph system to connect the south east with Darwin which in turn was connected by underwater cable to Asia and Europe. This allowed merchants in Sydney and Melbourne to negotiate prices with buyers in London prior to the despatch of cargoes across half the world. The cost of the construction and maintenance of the telegraph was considerable but so were the benefits. To the merchants of Australia the information conveyed by the telegraph was as tangible and real as the agricultural produce they sold, the ships that transported their produce and the energy that propelled them. But although sheep can be counted and energy measured, what exactly is information, how is it to be defined and what are the units by which it is to be measured? These questions were not answered until Shannon produced a mathematical theory of information in 1948. In the meantime a number of thinkers contributed to the development of ideas in this new field.

In 1927 Heisenberg proposed the principle of uncertainty. This revolutionised physics and philosophy. Newtonian physics was built on the idea that a cause led inevitably to a specific effect. The uncertainty principle proposed that a cause leads to a specific effect with a probability that is always less than 100%. We cannot predict the future with certainty. The uncertainty principle is important in physics but it is also important to information theory, in fact information and uncertainty are two sides of the same coin; information is the reduction of uncertainty; uncertainty is the reduction of information.

An important figure in the development of information was the Cambridge mathematician Alan Turing. In the 1930s he worked out the principles of a universal calculating machine. In the Second World War he was involved in breaking the enigma code, an achievement that made a major contribution to the war effort and demonstrated that in warfare information is as real and valuable as guns and tanks and aeroplanes. After the war Turing became a lecturer in mathematics at Manchester University and contributed to the development of the first digital computer.

In 1948 Shannon published a mathematical theory of information. Information was measured as the reduction of uncertainty on a log scale. Thus one bit of information reduces uncertainty by one half, two bits reduce uncertainty to one quarter of its initial value, three bits reduce uncertainty to one eighth and so on.

In 1953 Watson and Crick showed that DNA is the molecule of information. This ushered in the molecular biological revolution. Crick announced in the Eagle pub in Cambridge, to anybody who would listen, that he had discovered the secret of life - specifically how life begets life. This was not an idle boast; the structure of DNA revealed how information is stored and transmitted. The work of Watson and Crick is the basis of modern biology and is as famous and significant as Einstein's equation.

We are now surrounded by the artefacts of the age of information; word processing, spread sheets, e-mail, the internet, digital photography, telephone and television; but the theoretical basis of this new age is less well known.

DNA - the Molecule of Information

DNA is deoxyribonucleic acid. It is composed of nucleotides each of which consists of a sugar, a base and a phosphate group. In DNA the sugar is deoxyribose. Those who have not studied chemistry should not be put off by these terms, just regard them as names for different chemical building blocks. DNA is a double helix. The helical backbone is a long line of nucleotides with the bases pointing inwards. There are four bases in DNA; the pyrimidines, cytosine and thiamine, and the purines, adenine and guanine.

The double helix is two stranded; each strand is a long line of alternating phosphate sugar groups with the bases at right angles to the helical backbone. The key property in terms of information is that the base adenine always bonds with the base thiamine and guanine bonds with cytosine. Thus the double helix is like a helical staircase with steps formed by complementary bases. The steps are always adenine- thiamine or guanine-cytosine and never any other combination. If the two strands are separated DNA nucleotides can be added one at a time to the single strands to form two new double helices. These helices will be identical to each other and identical to the parent helix because in building the new helices adenine always pairs with thiamine and guanine always pairs with cytosine. In this way a double helix can replicate, creating two identical copies with information content intact.

The information in DNA is a linear alphabet of bases. There are four bases; ATGC. If the bases are taken two at a time there are 16 possible combinations i.e. AA, AT, AG, AC, TA, TT, TG, TC, GA, GT, GG, GC, CA, CT, CG, CC. If the bases are taken three at a time there are 64 possible combinations. Sixteen start with A i.e. AAA, AAT, AAG, AAC, ATA, ATT, ATG, ATC, AGA, AGT, AGG, AGC, ACA, ACT, ACG, ACC. A further sixteen start with T i.e. TAA, TAT, TAG, TAC, TTA, TTT, TTG,

TTC, TGA, TGT, TGG, TGC, TCA, TCT, TCG, TCC. Sixteen start with G and sixteen with C making 64 in all. These 64 triplets code for 20 amino acids, which are the building blocks of proteins. Thus a linear sequence of bases codes for a linear sequence of amino acids which in turn form a specific protein. Biological organisms are composed of a multitude of different proteins which act together to create life. The primary information for that complexity is stored in DNA.

The translation of information from DNA to proteins is slightly more complicated in that there is an intermediary which is RNA. RNA is ribonucleic acid. It is a linear molecule of nucleotides in which the sugar is ribose rather than deoxyribose. RNA also has four bases but the base uracil replaces thymine so that its constituents are cytosine, uracil, adenine and guanine. The transfer of information involves separation of the strands of the double helix and then on one strand a complementary RNA is synthesised so that the sequence ATGC in DNA pairs with UACG in RNA. The RNA molecule then carries the information specifying a protein from the nucleus to the cytoplasm of the cell where the protein is synthesised.

A gene is a linear sequence of DNA bases which specify a protein. The human genome has 25,000 genes in two sets making 50,000 genes in all. One set is inherited from mother and one from father. Each gene has approximately 1000 bases coding for a protein. Thus the number of bases coding for proteins in the genome is of the order of 50,000,000. The total number of bases in the genome, however, is over a billion and only a small fraction actually code for proteins. The function of the rest is not clear. Each gene consists of a series of exons and introns. The exons code for proteins but the introns are non-coding. Further there are regulatory sequences of bases before and after each gene concerned with switching the gene on and off. The situation is even more complicated in that each gene can code for more than one protein. There are 25,000 genes in the human genome but the number of proteins specified by these genes is in excess of 250,000 and could be as many as 500,000. Each gene is made up of a sequence of exons and the exons can be combined in various ways to produce a number of different proteins. This is achieved by copying DNA exons to RNA and then differentially splicing the RNA to produce a number of different RNA sequences which are in turn copied to form different amino acid sequences. Furthermore the proteins can be modified by enzymes (which are themselves proteins), adding on or removing chemical groups to produce different proteins. The information in the genome is therefore multiplied many times to produce the complexity of life.

Human cells have 46 chromosomes, 23 from mother and 23 from father. There are 22 pairs of autosomal chromosomes which are the same in males and females and one pair of sex chromosomes. Males inherit a Y sex chromosome from father and an X sex chromosome from mother. Females inherit an X sex chromosome from both parents. Thus males have 22 pairs of autosomal chromosomes and XY sex chromosomes. Females have 22 pairs of autosomal chromosomes and XX sex chromosomes. Each chromosome is a long double helix of DNA containing very many genes and long non-coding sequences. When cells replicate the double helices separate and a series of enzymes are used to synthesise complementary strands of DNA so that one chromosome becomes two identical chromosomes. At cell division the two sets of identical chromosomes go to opposite parts of the cell, a new cell wall is built dividing the cell in two and two identical cells are produced.

In this process billions of nucleotides are assembled in lines to faithfully copy the information held on the template molecule. The problem is that no biological process is completely error free and in the copying process mistakes can occur. There are a series of enzymes controlling the process and many of the mistakes that do occur are corrected, but a few escape detection. Uncorrected errors occur at a rate of approximately five in ten billion per base per cell division. Since there are 1000 coding bases per gene the chance that one of them is changed (mutation) is five in ten million per gene per cell division (five in ten billion times 1000 is five in ten million). A change in base sequence usually leads to a change in amino acid sequence and thus a new and often defective protein. This is a genetic mutation.

The key properties of DNA are:

- Information is stored in a linear sequence of bases: four bases taken three at a time gives 64 triplets that code for 20 amino acids.
- The human genome has 25,000 genes in two sets (one paternal, one maternal) but they code for up to 500,000 different proteins.

- Two identical copies are produced when the double helix replicates, this is how the information for life is copied from one generation to the next.
- The process of copying is not entirely error free and the mutation rate for genes is of the order of five in ten million per cell division.

The Principles of Information

Alan Turing was one of the architects of information theory. He did pioneering work in the 1930s on the mathematical and logical basis of computation. In the mid 1940s he was involved in the team that broke the enigma code; and in the late 1940s, when Shannon was developing the mathematical theory of information, Turing and colleagues built the first digital computer in Manchester, UK.

Turing was a mathematician but he was also interested in the brain. He thought the mind was a machine and he wanted to produce a machine that could think; hence the digital computer. It is probably for this reason that experimental psychologists were the first group of biologists to be interested in information theory. They did not make the mistake of regarding the new digital computers as equivalent to the brain, because in many respects they were not. Instead they argued that digital computers process information and the brain processes information so that the general principles that apply to all information systems will apply to both. Thus the principles of information processing, exemplified by computers, will help in understanding the way in which the brain works.

General principles that apply to all information processing systems are as follows:

- All information processing systems have a finite capacity.
- Information is processed in a background of noise and therefore there is always a finite possibility of error.
- Information processing systems decay with time according to the laws of entropy and therefore the error rate will rise with time.

- All complex information processing systems require a high degree of redundancy in order to reduce error to a minimum.

These principles will be used throughout this book and it is important that they are understood.

The first principle is self evident; it is a statement of the obvious. The amount of information that can be transmitted along a telegraph wire or along a telephone wire is limited by the capacity of the system. In the same way the amount of information that a computer can store or the brain can process in unit time is limited. The fact that a statement is self evident or obvious, however, does not mean it is not useful; this is known to mathematicians and physicists but perhaps less well appreciated by biologists.

The second principle is of fundamental importance. All matter consists of chemicals in which electrons are distributed around nuclei consisting of protons and neutrons. The electrons have negative charge and the protons positive charge. The neutrons are neutral. The precise position and motion of an electron cannot be specified with certainty. It has a likely position but its actual position is subject to random fluctuation. This is Heisenberg's uncertainty principle. The transmission of information in a physical system, such as along wires in telegraphy or telephony, involves the motion of electrons which are subject to random fluctuations; thus there is always a possibility that information is lost and errors occur. Noise is random fluctuation and it cannot be completely eliminated, therefore errors cannot be completely eliminated.

The second law of thermodynamics is that entropy increases with time. Entropy is a measure of disorder and the law states that with time systems become more disordered, more random. If information systems become less ordered, less carefully regulated, less well controlled with time then the error rate will rise. This is a universal principle, all machines wear out, components cease to function, noise increases and deterioration in performance is inevitable.

The fourth principle requires the most explanation and there is a section below devoted to this important concept. In a complicated information processing system there is a finite chance that a component will not work.

Information Theory

As systems become more complicated, with more components to break down, the risk of failure increases. A way to avoid failure is to create a highly redundant system so that if one component fails there is another to take over the task. The more complex the system, the higher the level of redundancy required.

It is easy to see that the brain processes information and so the above general principles can be applied to psychology. Information flows along neurones in the brain, as a disturbance of electrical charge, just as information flows along telegraph wires. Memories, information of past events, are stored in the brain just as information is stored on computer; the precise physical form might differ but the principles are the same. The capacity is finite; the process of transmission and storage is subject to error, the systems decay with age but redundancy reduces error and slows the deleterious effects of temporal change.

What is perhaps less obvious is that other aspects of biological function can be analysed in terms of information processing and information theory. When an infant is born it is quickly colonised by a multitude of different bacteria which cover external and internal body surfaces such as the skin, the upper respiratory tract and the gastrointestinal tract. Indeed there are more bacteria on the body surface than there are human cells in the body. We are, at least in numerical terms, mainly bacterial and only partly human. The immune system has the task of protecting us against these bacteria. Most are relatively harmless but some are potentially lethal. Bacteria consist of a number of proteins specified by genes, just as human cells consist of proteins specified by genes. In the case of bacteria there are approximately 5000 genes specifying approximately 5000 proteins (compare this with human cells in which 25,000 genes specify up to 500,000 proteins). The immune system must classify these proteins as similar to self or different from self. It is important to avoid attacking similar to self proteins as that could lead to an attack on one's own cells, causing autoimmune disease. Equally it is important to recognise different from self proteins and attack them if they are carried by potentially lethal organisms. A further complication is that there are more different from self proteins than there are immune cells (lymphocytes) to recognise and attack them. Thus the immune system must concentrate on the few different from self proteins carried by potentially lethal bacteria and avoid an attack on different from self proteins carried by harmless bacteria. This complex task is a problem in information processing, the proteins must be recognised and classified and the information stored for further use. The general principles of information processing will apply: the system has a finite capacity, the information is analysed in noise and errors will occur, the system will deteriorate with age, and if we look carefully there will be evidence of redundancy.

The brain processes information and the immune system processes information; but the same also applies to all physiological systems. The beat of the heart, the control of blood pressure and respiration, all involve the flow of information along nerves and the release of chemical messengers in to the blood. Indeed we can only be regarded as individuals because the billions of cells of our bodies communicate and act in concert to maintain our integrity. We are processors of information and the principles of information theory can be used to understand the way in which we work.

Redundancy

The dictionary definition of redundancy is "superfluous, superabundant or in excess"; this is the way that the word is used in common parlance and it has negative connotations. At one time employees were sacked; now they are made redundant, but the situation is no less unpleasant. Management consultants trim the fat, reduce redundancy in the system, improve efficiency and push up profits; or at least that is what they claim.

Information theory leads us to an opposite conclusion. Redundancy is good; indeed redundancy is essential for the safe and effective function of a complex system. If a message is sent along a noisy communication channel part of the message will be lost; if there is no redundancy in the message the recipient will not know what the sender wished her to know and communication will fail. Since all communication channels are noisy to some degree, sending non-redundant messages is not a good or efficient strategy.

My telephone number is a non-redundant sequence; if you dial it and get one number wrong you will not get through. Language by comparison is a highly redundant mode of communication and even if there are a number of errors in transmission the message usually survives. Consider the instruction "turn left at the next junction", the following variations would be understood: "tern lft at nixt junction", "tn left at nxt jnction", "tjrn left at nejt unction" etc. All languages have this property, they are highly redundant in both spoken and written form and as a result language is an effective mode of communication.

Complex systems have many component parts and there is always a finite chance that a part will fail or an error will occur. The possibility of failure can be reduced if there are backup parts so that if one fails there is another to perform the task; this is redundancy. As complexity is increased and the number of component parts is increased the likelihood of failure also increases and some degree of redundancy becomes essential. The more complex the system, the more important the task, the higher the level of redundancy required. Redundancy is a positive not a negative attribute.

I am a consultant histopathologist. I report biopsy samples removed by surgeons at operation. A thin section of the sample is cut and stained and I examine it down the microscope. The majority of the lesions removed by surgeons are relatively harmless (benign) but a few are potentially lethal (malignant); it is obviously important that I, as a histopathologist, distinguish accurately between the two. Studies indicate that histopathologists are right 99% of the time but make mistakes in approximately 1% of reports. These are mistakes that will significantly affect patient management. The problem is that I report over 100 samples per week and that means more than one significant error per week. If another histopathologist, who had the same error rate as me, checked my work then the chance that we both make the same mistake is not 1 in 100 but 1 in 100 times 1 in 100 which is 1 in 10,000. With this system mistakes would not occur at the rate of one per week but once in two years. If a third histopathologist checked the samples mistakes would then occur once in 200 years (1 in 100 times 1 in 100 times 1 in 100 equals 1 in one million). This would be a highly redundant system, and a very expensive system, but it would be effective in reducing errors.

Let us imagine an office with a female executive and seven male secretaries (all things are possible). The executive dictates a letter, one of the secretaries types the letter and then all seven secretaries read it to check

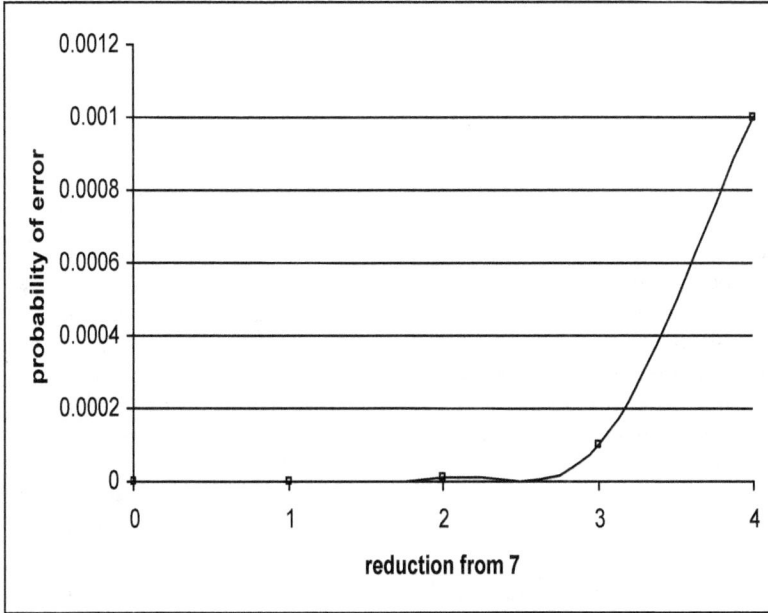

Figure 1.1

The probability of failing to detect a mistake as the number of secretaries checking a piece of prose is reduced from seven. The probability of one secretary failing to detect a mistake is 1 in 10, the probability of all seven secretaries failing to detect the mistake is 1 in 10 million. The probability of six secretaries failing to detect the mistake is 1 in a million, the probability of 5 secretaries failing to detect the mistake is 1 in 100,000 and so on. Removing one or two secretaries has little effect removing a third has an observable effect and a fourth has a larger effect. This is synergistic interaction and is a property of a redundant system.

for errors. The chance of an error remaining undetected after the letter has been read by one secretary is 1 in 10 (10% or 0.1). The chance of an error remaining undetected after the letter has been read by two secretaries is 1 in 100 (1% or 0.01). If all seven read the letter the chance of an undetected error is 1 in 10,000,000 (0.1 times 0.1 times 0.1 times 0.1 times 0.1 times 0.1 times 0.1 = 0.0000001). The error rate is shown in graphical form in figure 1.1. With seven secretaries the error rate is very low. If one secretary is sacked (those management consultants again) then the error rate will rise ten fold but in absolute terms the difference between 1 in 1,000,000 and 1 in 10,000,000 is negligible. If two secretaries are sacked the error rate becomes 1 in 100,000, but this is still a very small rise. Loss of a third secretary leads to a rise to 1 in 10,000 and a fourth brings the rate to 1 in 1000. The graph shows initially a very small rise in error rate as one or two secretaries are lost but then a much more marked rise as a third and a fourth is lost. If the reputation, and the income, of the executive depended on producing error free letters and reports she could afford to lose one or two secretaries but further loss would be harmful. This is an important property of a redundant system and it is found throughout biology. A complex system requires sufficient redundancy to lead to a very low error rate, if one or two components are lost the change will be small and in practice not noticeable or measurable; but the loss of one or two more components will have a noticeable effect and further loss is incompatible with survival.

The central message of this section is that complicated systems need a high level of redundancy. The most complex systems are living systems, and the most complex of all are human beings. Thus whenever we examine human physiology (function), morphology (structure) and pathology (disease) we find redundancy. When bacteria attempt to invade the body and cause disease there are a whole series of defences. Bacteria need to adhere to the surface prior to invasion and this is prevented by the secretion of mucus by human cells which carries the bacteria away. If the bacteria can swim through the mucus to attach to the cells that secrete it then another defence is the secretion of a molecule IgA which combines with the bacteria and stops them adhering to the surface cells. If the bacteria avoid the IgA molecules and invade through the surface cells into the body they are then attacked by a complex array of killer cells (neutrophils, eosinophils and macrophages) aided by proteins from the blood including IgG, IgE, complement and cytokines.

The immunoglobulin molecules IgA, IgG and IgE are not one compound but thousands of different compounds that recognise specific proteins on the surface of bacteria. If the appropriate immunoglobulin molecules are present they will coat the surface of the bacteria and allow neutrophils to engulf the bacteria and destroy them. The defences are a complex redundant system and on most occasions they succeed in preventing infection. If components are missing such as IgA or even neutrophils they still succeed most of the time but infection is more likely to occur.

The immunoglobulins IgA are secreted onto the mucosal surfaces of the respiratory and gastrointestinal tracts. The molecules coat the surface of specific bacteria and prevent them adhering to the mucosal surface. The mucus secretions of the mucosal cells then carry the bacteria away, they are escorted of the premises and thereby do no harm. The IgA molecules are directed to removing potentially harmful bacteria and leave the rest alone. There are individuals who are unable to manufacture IgA because of genetic mutations that they carry. These individuals live normal lives and do not appear to have a greatly increased risk of infection. At one time this led scientists to think that IgA had no significant function and was unnecessary, but it is now realised that IgA is probably the single most important defence against infection. More IgA is manufactured per day for instance than IgG; the latter guards against infection in the body and is essential to life. At first sight it appears a paradox that IgA is an important part of an essential system yet we can survive without it; but it is not a paradox, it is the essence of redundancy. The more important the task the more important it is that any one component is "superfluous, present in excess".

The brain is also a highly redundant structure. We steadily lose brain cells throughout life but our brains function adequately into old age. Redundancy in the brain allows us to age gracefully. With advancing years we become progressively less good at solving new problems and are less willing and able to take on new ideas; but we are good at solving old problems because we have met them before, and we are quick to recognise bad ideas purporting to be new. Thus as we get older we lose intelligence but gain wisdom; redundancy underlies both.

Genetic Mutation

The moment of human conception occurs when a spermatozoon (singular of spermatozoa), carrying 25,000 genes from father, penetrates and fuses with an oocyte, carrying 25,000 genes from mother, to form a zygote. The zygote is a single cell, it has 50,000 genes but the two sets of 25,000 are virtually identical (but not completely see below). The zygote divides by mitosis to form two cells. The process of mitosis was described above. The double helix of the chromosome divides into two and on each strand a complementary strand of DNA is built up so that one double helix becomes two identical double helices. One of the double helices goes to one half of the cell, the other goes to the opposite half and the cell then divides into two. In this way 46 chromosomes replicate to produce two identical sets of 46 chromosomes, one set in each of the two new cells. There is, however, the possibility of mutation occurring during the copying process of mitosis. The mutation rate is such that a single gene has a one in a million or one in ten million chance of mutation occurring per cell division (see above).

The zygote divides to form two cells; two cells divide to form four cells; four cells divide to form eight cells and so on. If this process is repeated 30 times (30 cell generations) a billion cells are produced weighing approximately one gram. Forty cell generations produces a thousand billion cells weighing one kilogram. The adult male secretary weighs 70 kilograms and the female executive weighs a little, but not a lot, less. In thirty or forty or fifty cell generations any one gene has thirty or forty or fifty chances of undergoing mutation.

Adults are composed of many billions of cells each containing 46 chromosomes in two sets of 23, one from each parent. Spermatozoa and oocytes (the male and female gametes), however, contain just one set of 23 chromosomes but each chromosome is a mixture of maternal and paternal genes (i.e. a mixture of genes from the mother and father of the adult who has produced the gametes). A reduction from 46 chromosomes to 23 chromosomes occurs in meiosis. In this process the two sets of 23 chromosomes align and then random breaks occur along the length of the chromosomes with rejoining between adjacent chromosomes. The result is each chromosome contains a mixture of maternal and paternal parts. The newly formed chromosomes then pull apart and one from each pair goes to the opposite end of the cell. The cell divides and two gametes are formed each with one set of 23 chromosomes. The term haploid is used for gametes that have one set of genes and the term diploid for cells that have two sets of genes.

In fact the process in humans is slightly more complicated: the chromosomes first replicate and then undergo reduction division so that one diploid cell forms four haploid cells in meiosis.

Oocytes are approximately 35 cell generations from the zygote and spermatozoa are over 50 cell generations form the zygote. There are 50,000 genes per cell, the new mutation rate is between 1 in 1,000,000 and 1 in 10,000,000 per gene per cell division, and zygotes are 35 cell generations from the maternal zygote and over 50 from the paternal zygote: it is simple to calculate from this that on average there will be approximately one new mutation in a zygote that was not present in either of the parental zygotes. Let us designate this quantity as P, it is the new mutation rate per human generation.

There are a number of ways in which mutations occur when cells divide and DNA is copied. They all involve changes in the sequence of bases in DNA. If the mutation affects that part of the DNA that codes for proteins then the sequence of amino acids in the protein will change. Sometimes the new protein works as well as the old protein, this is a neutral mutation. More commonly the new protein does not work at all, this is a deleterious mutation.

Neutral Mutations

There is no disadvantage to neutral mutations. They are passed on from generation to generation and slowly increase in number as new neutral mutations arise. Genetic differences between individuals are mainly due to the different neutral mutations that they have inherited. The major differences between races are due to neutral mutations; groups of individuals who inter-marry will share a common pool of neutral changes that will be absent from groups with whom they do not inter-marry.

One set of neutral mutations involve proteins that are concerned with fighting infection. There are proteins on the surface of our white blood cells that are instrumental in recognising and attacking specific bacteria and viruses. One version of the surface protein might be good at dealing with bacteria A, but not so good with bacteria B. Another version of the same protein is good with B but not with A. If bacteria A is prevalent the former mutation is advantageous and the latter disadvantageous, but when times change and bacteria B is prevalent then the situation is reversed. Thus a specific neutral mutation can be advantageous in one setting but disadvantageous in another, overall it is neutral.

Neutral mutations are important in susceptibility to infectious and autoimmune disease. Bacteria and viruses and other micro-organisms have evolved numerous different strategies for invading the body, avoiding defences and causing disease. We have equally complex systems for fighting and preventing disease. Neutral mutations do not impair that ability overall, but they can, as noted above, improve one part at the expense of another. Thus if we investigate one specific bacterial or viral infection it is common to find that certain genes confer susceptibility. Autoimmune disease occurs when white cells designed to fight infection turn instead and attack our own cells. The reason for this is not fully understood but one possible explanation is based on molecular mimicry. One of the tasks of the immune recognition system is to distinguish proteins on bacteria that do not resemble self proteins from those that do and attack the former but not the latter. Autoimmune disease, according to this concept, occurs when there is an error in this recognition process and white cells attack a protein on bacteria which resembles one of our own proteins, they destroy the bacteria but also attack the same protein on our own cells. Studies show that certain neutral mutations confer susceptibility to specific autoimmune diseases but equally protect against others. It appears that neutral changes in some way interfere with recognition processes but only for specific organisms not all organisms.

Deleterious Mutations

A deleterious mutation is a change in a gene which leads to functional loss. If P (the rate at which new deleterious mutations enter the human genome per human generation) is 1 or above then the number of deleterious mutations will progressively rise. Unchecked or unopposed this would lead to extinction of the human race. There must be some process that opposes this steady rise.

It is possible to estimate the number of deleterious mutations in the genome from the frequency of recessive disease in the progeny of cousin marriages. Let us define a quantity Q that is the average (mean) number of deleterious mutations that were present in the zygote of individuals who have survived to adult life. Most genes come in pairs, one from mother and one from father. If one gene of a pair is deleted the other can produce the appropriate protein and there is usually no observable disease. If by chance both genes are deleted then the protein is absent and this leads to recognisable recessive disease in many, but not all, cases. Consider a single deleterious mutation in a grandfather or grandmother, the chance that the same mutation is present in a grandchild is 1 in 4. If two grandchildren marry (cousin marriage) the chance that they both carry the specific mutation is 1 in 16. The subsequent chance of a child inheriting the mutation from both parents is 1 in 4 giving a risk of recessive disease in a child of a cousin marriage as 1 in 64 for any single deleterious mutation carried by the grandparents. The grandparents each have Q deleterious mutations on average therefore there are 2Q deleterious mutations that can be passed from grandparents to cousins. The risk of recessive disease not arising for any one mutation is 63/64, therefore the risk of it not arising for any one of the 2Q genes is (63/64) raised to the power 2Q. Those not used to probability should remember that the chance of getting heads when a coin is tossed is 1 in 2 (1/2 or 0.5), the probability of not getting heads on two consecutive occasions is 1/2 times 1/2 which equals 1/4. The probability of not getting heads on four consecutive throws is 1/2 times 1/2 times 1/2 times 1/2 which is 1/2 raised to the power 4 i.e. 1/16. The probability of not getting heads on Q occasions is 1/2 raised to the power Q.

The actual calculation is more complex because one has to allow for P (the number of deleterious mutations that enter the human genome per generation), for the possibility that recessive disease could lead to abortion and that some recessive disease might not be diagnosed. Studies in parts of the world were cousin marriage is common, such as the Middle East, show that recessive disease affects only a small minority of children and calculations indicate that Q is less than 10 and probably closer to 6 or 8.

If P is greater than 1 and Q is less than 10 there must be some process of selection that prevents the build up of deleterious mutations in the genome.

The Process of Selection

Genes code for highly redundant highly complex biological systems. A single deleted gene will lead to loss of its protein product and interfere with some component of a redundant process; but if the system is highly redundant it will have no observable or measurable effect and will not lead to selective disadvantage. In this way it is possible for a number of deleterious mutations to accumulate in the genome with no harmful effect. There will be no selection against single deletions in redundant systems. As Q increases, however, there is an increasing chance that a single system has two or three or more deletions. If this occurs then function will be compromised and selection can operate.

This process has been studied in computer models. Consider a single unicellular organism with a haploid genome. The term unicellular is self evident it means a single cell; bacteria for instance are unicellular. The term haploid means one set of genes rather than two sets (diploid) as occurs in most human cells. This unicellular organism, hereafter referred to as Organism A, is attacked by a parasite and defends itself with a highly redundant system specified by genes. The system has m components each specified by y genes. The probability that a single parasite avoids any one component is p; if the parasite avoids all m components then organism A is killed. While it feeds organism A is attacked by q parasites; if it successfully defends itself against all q parasites then it divides by mitosis to produce two daughter cells. During mitosis there is a probability b that one of the daughter cells acquires a deleterious mutation that will effectively delete one of the m components.

If the computer model is run using reasonable values for the various variables m, p, q and b, then the number of cells of organism A increases with successive divisions to reach vast numbers. In the process the cells accumulate deleterious mutations and eventually there are no cells left with

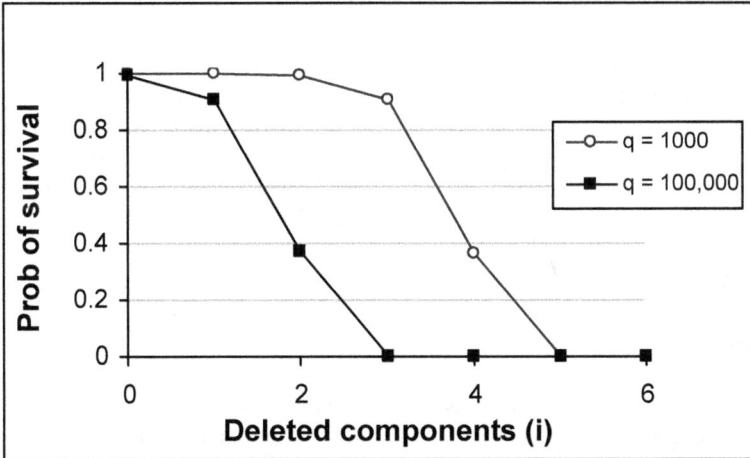

Figure 1.2

The probability of survival of a unicellular organism which is attacked by q parasites. Loss of one or two genes by mutation has a small effect when q = 1000 but loss of subsequent genes has a much larger effect. This is another example of synergistic interaction in a redundant system. The model also shows that if q is increased the level of redundancy is reduced and even the loss of one gene has a noticeable effect.

[In this model there are i deleted components from the system that fights the parasite, and j deleted genes. One deleted gene deletes a component but some components can have more than one deleted gene; therefore i does not necessarily equal j. The graph shows i not j.]

all m components functioning. If $m=7$ for instance and $b=0.1$ or 0.01 then eventually all cells with 7 or 6 functioning components disappear and those remaining have 5 or 4 or even 3. The organism is still successful in that it continues to grow and increase in numbers but it has lost the very high levels of redundancy that it had initially. If the environment changes and becomes more threatening, which is modelled by increasing the value of q, then organism would be extinguished. The reason is that it has lost redundancy when it was subject to a low level parasitic attack (low q) but is then destroyed by a higher level parasitic attack (fig 1.2).

If $b = 0.001$, the level found in bacteria, then a few cells will maintain redundancy over the long term and will be able to survive if the environment becomes more harsh. However, if we attempt to construct a more complicated organism by increasing the number of genes in the genome we automatically increase the value of b and the organism cannot survive long term in a changing environment. There is therefore a limit to the complexity of asexual organisms.

Let us now introduce sex into the model. Organism E is similar to organism A in all respects except that it engages in occasional bouts of sexual reproduction. In this process two cells fuse at random to form a diploid cell (two sets of genes). The fused cell then undergoes meiosis with a high degree of recombination (the process in which the chromosomes break and cross over). This ensures that the genes from the parent cells are maximally mixed in the daughter cells. If the two cells that fuse each have 2 mutations the daughter cells could have 0,1,2,3, or 4 mutations. The difference between asexual and sexual reproduction is that in the asexual process the daughter cells always have the same number or more deleterious mutations than the parent cell, never less: but in sexual reproduction the daughter cells can have fewer mutations than the parent cells, although the mean number is the same. When organism E is run in the model there are always a few cells with no deleterious mutations and therefore the progressive build up of mutations and loss of redundancy is avoided. Figure 1.3 shows what happens when organism A and organism E compete in a closed environment for finite resources. This figure tells you everything you need to know about sex. Is it worth all that trouble, all that fuss, all that heartache, all that mess? The answer is yes. Organism E outgrows and replaces organism A. Initially organism A has a growth advantage this is

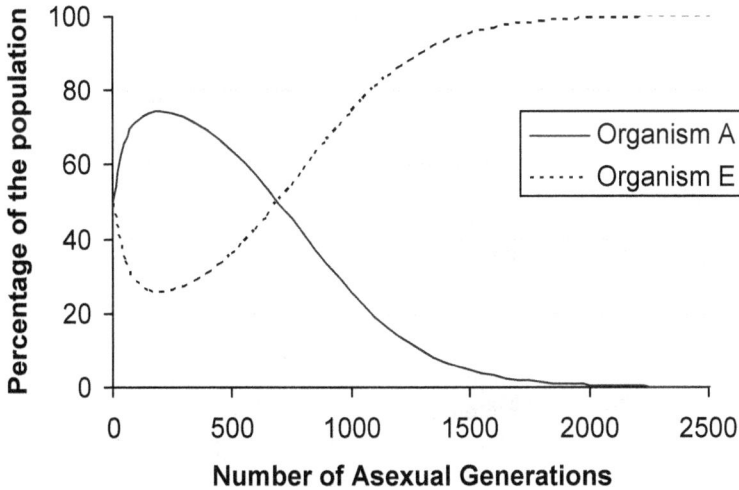

Figure 1.3:

Organism A reproduces asexually and organism E reproduces sexually. The graph shows that if they compete in an environment with finite resources organism A has an initial advantage but it loses out in the long term. Organism E preserves redundancy but organism A gradually loses redundancy.

because the sexual process distributes deleterious mutations at random to progeny and actually creates disease in that some of the progeny gain several mutations and quickly die out. In the longer term, however, the sexual process is advantageous. The asexual form gradually accumulates mutations and its growth rate slows; the sexual organism maintains its growth rate and slowly overtakes the asexual form. Sexual reproduction conserves redundancy and thereby allows an organism to survive in the long term.

Increasing complexity by increasing the size of the genome creates a problem for both sexual and asexual organisms; a bigger genome leads to an increased chance of mutation. This is a major problem for asexual organisms in that mutations accumulate rapidly and extinction is inevitable. It is also a problem for sexual organisms in that it slows the growth rate. One way to overcome this is to use existing genes in new combinations to perform new tasks; thus complexity increases but the number of genes does not increase as fast and therefore the mutation rate does not become too high. Recent work has shown that bacteria have 5000 genes, yeast have 6000, flies have 13,000, worms have 18,000, plants have 26,000, mice have 25000 and men have 25000. We are more than 6 times as complex as bacteria, or at least we think we are, and therefore it is tempting to accept the idea that our genome is more complex because we use our existing genes to do several tasks. Equally evolution from mice to men is not due to the creation of new genes but must be due to using old genes in new ways. These ideas are particularly important when we come to consider the genetic specification of sexual attraction and physical beauty. Put simply genes can code for beauty and the same genes can code for function so that one can be a marker for the other.

These various ideas can now be put together to examine selection against deleterious mutations in humans. Most recent evidence indicates that P is more than one but less than two, whilst Q is less than 10 and perhaps close to 6 or 8. Spermatozoa and oocytes are gametes. During meiosis deleterious mutations are distributed at random from the parent cell to the gametes. The gametes then fuse at random to form zygotes. The mean number of deleterious mutations in zygotes will be P + Q (this is the same as the parent cells which started with Q but have accumulated an extra P). The actual number in any one zygote will vary as shown in figure 1.4. Zygotes with only a few deleterious mutations will have high levels of

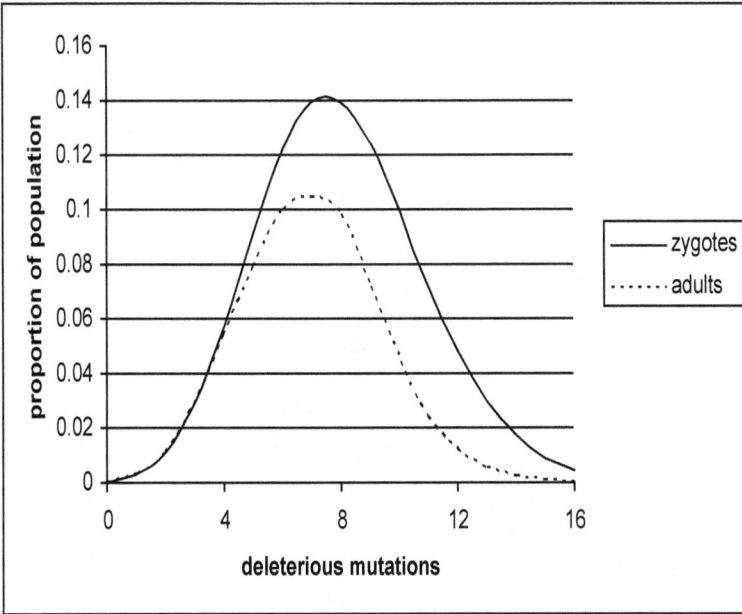

Figure 1.4

The figure shows a Poisson distribution of deleterious mutation in zygotes with a mean of 8. The probability of survival of zygotes decreases as the number of deleterious mutations increases. The resulting distribution in the germ line of zygotes which survive to become adults is a skewed Poisson distribution with a mean of 7. The population is in balance if P = 1.

redundancy in all systems; they are likely to survive, to develop into adults, to marry and to have children. Zygotes at the opposite end of the distribution are more likely to have at least one system with several deleterious mutations, that system is likely to fail and the zygote will abort. Figure 1.4 shows the result if there is a gradient of survival from low at the right to high at the left. The distribution of deleterious mutations is shown in the survivors i.e. those who survive to become adults. The mean of the distribution is Q and the population is in balance. P mutations arise in each generation but are lost during selection.

The models indicate that approximately 50% of the zygotes that form do not proceed to term. This fits reasonably well with the observation that the probability of a pregnancy arising and going to term in any one menstrual cycle for a couple engaging in regular unprotected sexual intercourse is 1 in 4. Regular intercourse in this context is 2 to 3 times per week. This fits with a zygote forming 50% of the time and surviving 50% of the time.

The distribution of deleterious mutations in those who survive to become adults also shows marked variation. Those with few deleterious mutations are likely to have systems that function optimally, but those at the upper end of the distribution will suffer some impairment of function. Thus sexual reproduction conserves redundancy but in the process produces a degree of inequality.

Conclusion

This chapter has covered a lot of material which will be unfamiliar to many readers. I am not going to attempt to summarise the chapter but I do recommend that the various sections in chapter one are revisited as the other chapters are read.

Chapter Two

Intelligence

Intelligence Tests

In 1904 the Ministry of Public Instruction in Paris established a commission to identify children in need of remedial education. Alfred Binet developed a test to measure objectively the intellectual ability of students with learning difficulty. This test was the forerunner of standardised intelligence tests. The test measure was the intelligence quotient (IQ) defined as mental age divided by chronological age multiplied by 100. A child aged 8 years who could perform at the level of the average 10 years old had an IQ of 10/8 times 100 = 125. A child of 10 years who performed at the level of a child of 8 years had an IQ of 8/10 x 100 = 80.

This definition of IQ could not be used for older children and adults. Instead the IQ is now standardised so that the mean is 100 and the standard deviation is 15 (figure 2.1). Scores obtained on any one test are converted to this scale. For the general population 34% have an IQ between 100 and 115, 14% have an IQ between 115 and 130, and 2% have an IQ over 130. The distribution is symmetrical around 100 so that the same percentages apply below 100 i.e. 34% between 85 and 100, 14% between 70 and 85 and 2% below 70. It should be appreciated that the standardisation process creates the symmetrical normal distribution shown in figure 2.1. The scores obtained are not necessarily normally distributed, although they tend to be close to a normal distribution. Furthermore we should not assume that the distribution of intelligence and general ability is normal and symmetrical.

There are a number of different intelligence tests and they measure a number of different factors. The Washington Pre-College Test Battery, for instance includes the following:

- Reading comprehension: Answer questions about a paragraph
- Vocabulary: Choose synonyms for a word.
- Grammar: Identify correct and incorrect grammatical constructions.
- Quantitative skills: read word problems and decide if they can be solved
- Mechanical reasoning: Examine a diagram and answer questions about it.

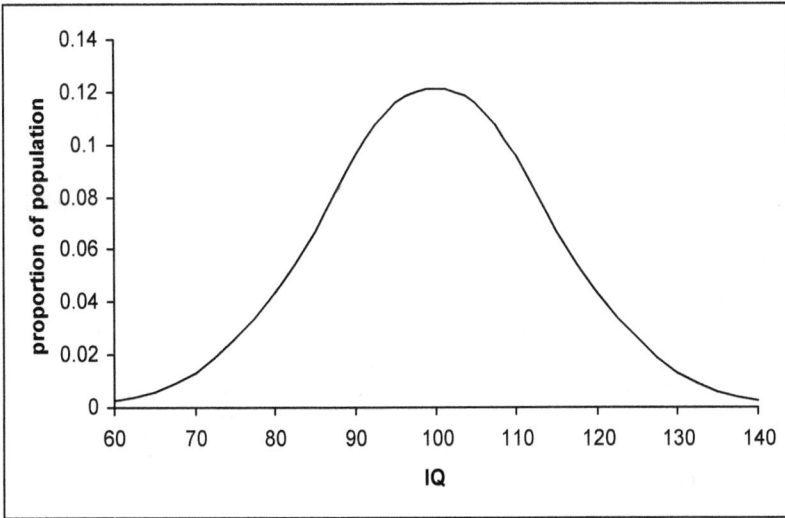

Figure 2.1

The IQ is normally distributed with a mean of 100 and a standard deviation of 15. 68% of the population have IQ scores between 85 and 115, 16% have IQ scores over 115, and 2% have a score over 130.

- Spatial reasoning: Indicate how 2-dimensional figures will appear when folded.
- Mathematics: A test of high school algebra.

In practice it is found that those who do well on any one test tend to do well on the others. Expressed in statistical terms the results on the tests are positively correlated. Consider a class of 30 individuals who all take the above tests. The class are placed in rank order from first to last for reading comprehension and for mathematics. If the order is exactly the same for the two tests then the correlation coefficient is 1. The individual who is first (rank one) in reading is also rank one in mathematics, the same person is second in both, the same individual is third in both and so on to the unfortunate person who is bottom in both tests. If the order were reversed in one compared with the other the correlation coefficient would be -1. In this case the person who was first in reading would be last in mathematics; the individual next to the top in reading would be next to the bottom in mathematics and so on. If there were no correlation between the two lists then the correlation coefficient would be 0. In practice the correlation coefficients between the tests are always positive and vary between 0.3 and 0.7. It is because the scores on the various tests correlate positively that we can talk about intelligence as a unitary concept; but we should also bear in mind that the tests measure different abilities and the degree of correlation is always less than 1.

An enormous amount of research has been carried out on intelligence tests since the work of Binet in 1904. The tests have proved to be both useful and controversial. They are most useful when a large number of individuals from the general population have to be assessed as for school or college entry or deployment in the armed forces. They cause most controversy when they identify differences between groups and the spectre of genetic superiority and inferiority is raised. Over a century of research has, however, led to some generally accepted facts and in this chapter I intend to briefly present those facts and then see how they can be explained using the ideas presented in chapter one.

Key Facts on Intelligence and the IQ

In the discussion below intelligence is simply defined as that which is measured by the IQ test.

- Performance on the various components of IQ tests is positively correlated. Those who do well on one part tend to do well, or are at least above average, on the other parts. The tests therefore are internally consistent.
- Performance on an IQ test correlates positively with scholastic and academic success. Those who do well in the IQ tests tend to do well in school examinations, are more likely to stay on at school, are more likely to go to College, are more likely to gain a degree and are more likely to pursue an academic career.
- Performance on IQ tests also correlates positively with success in life in the most general sense of that term. Those with an above average IQ are more likely to gain a high status job; they are more likely to have an above average income; they are more likely to marry, to have children, to maintain their health, to live a long life and to stay out of prison.
- There is a genetic contribution to IQ. The mean IQ of children is close to the mean IQ of their parents and their siblings. Identical twins (monozygotic) form from a single zygote and have identical genes. They tend to have a very similar IQ even if reared apart. Non identical twins (dizygotic) are formed from two separate zygotes, they have 50% of their genes in common as do non twin siblings. The IQs of non identical twins are not as close as those of identical twins but they are equally as close as non twin siblings. If dizygotic twins are reared apart their IQs are less close than if they are reared together, but not as disparate as children in general.
- The twin studies also show that there is an environmental contribution to IQ. Identical twins do not have identical IQs, particularly if reared apart. Furthermore performance on IQ tests has increased over the last century by up to 15 points; this indicates an environmental effect as our genetic constitution changes more slowly.

- The most inflammatory aspect is that there are consistent differences in measured IQ between certain racial groups. In some reports the mean score for Blacks in the USA is 16 points below the mean score for Whites.

The above results are well known and well accepted but they are nonetheless extremely surprising and difficult to reconcile with conventional ideas. Why is there a positive correlation between different aspects of the IQ test? In general if an engineer is asked to design a machine to perform a number of different tasks there has to be a trade off between the varying requirements. Optimising the design to deal with task A tends to compromise the ability to perform task B. In this situation one would expect that performance on the various components would be negatively correlated. Consider asking an engineer to design a machine for transportation. He could produce an aeroplane, a ship, a train or a car. Which is the best? The question cannot be answered. The aeroplane is best if you want to travel across the world in a short time, but a car is best if you want to go to the shops. A ship is best if you want to transport a large cargo at low cost. These machines cannot be put in rank order from best to worst, they are equal but different. When we take human machines, however, we can rank them using the IQ for a whole range of different abilities. It is most odd.

We are surprised that performance on the components of IQ tests is positively correlated but we should be astounded that performance also correlates positively with success in life.

The suggestion is that those who are good at one thing are good at everything, even fighting disease and staying alive. Why? Those who pursue a special interest in their careers or in their hobbies know that by concentrating on one thing they can excel but it is at the expense of other interests and skills. You cannot be good at everything but the IQ results imply the opposite.

We should not be surprised that the environment and genetic constitution interact to determine IQ, because this is true of all things human. Our physiology, morphology, psychology and pathology are determined by nature and by nurture not one or the other in isolation. The way in which they succeed in determining intelligence is, however, not known. The most disturbing aspect of research into intelligence is the evidence of racial differences in measured IQ. Those who consider that IQ measures innate ability are led to the conclusion that some races are superior to others. Those who consider that the innate component is genetically determined will conclude that some races are genetically inferior. These are uncomfortable conclusions but fortunately they are wrong conclusions (see below).

A Clever Robot

Let us conduct a thought experiment, an activity beloved by theoretical physicists and despised, wrongly, by practical biologists. The experiment involves the design and construction of a clever robot; it can mow the lawn, do the housework, cook the dinner, play chess to a high standard, and perform double somersaults. The robot is controlled and specified by a large, complex and highly redundant set of instructions. It has neural networks that enable it to learn from the environment so that the housework and cooking can be improved by feedback from consumers and its chess playing improved by exposure to challenging opponents. It is undoubtedly a clever robot and we could construct a robot scale of ability, give it marks for its accomplishments and calculate an ability quotient - a robot IQ.

The robot IQ would depend on the quality of the instructions and on the quality of environmental experience. Scores on the chess component, for instance, would be influenced by the number of games the robot had played and the standard of opposition it had experienced.

The next stage of the experiment is to delete instructions at random from the control system. Since the instructions form a highly redundant system one or two deletions will have no observable or measurable effect. The robot IQ will not change. As the number of deletions is increased, however, subtle changes in ability will be noted; housework done reasonably well but not very well, lawn mowing not quite up to standard, chess playing less than brilliant. The measured robot IQ will start to fall. The decrease in performance will not be even across all abilities, some may be more markedly affected than others but all will suffer to some degree. If the process of deleting instructions is continued the decrease in ability will be more marked and obvious, eventually some components will be completely lost and the robot will cease to function.

The robot's cleverness was not due to any one instruction but to the sum total of all instructions acting together. The cleverness was removed, however, by deletion of only a small proportion of the instructions. The robot IQ was inversely related to the number of deletions.

The experiment is now repeated using 100 robots rather than one. Initially they are all identical in terms of instructions although the environment in which they work will vary. The robot IQ will be high for all robots initially but there will be some variation depending on experience. Deletions are now introduced at random with some having one, some two, some three and so on to a number that causes breakdown. There will be a positive correlation between performance on the various abilities for the set of 100 robots. Those with few deletions will be good at everything, those with a lot will be bad at everything. Most will be somewhere in the middle. The robot IQ will be inversely correlated with the number of deletions in individual robots.

Intelligence and Deleterious Mutations

Humans begin life as a single cell, the zygote, which contains two sets of 25,000 genes. The genes provide the information that guides development. In a facilitating environment the process will proceed with the minimum of error to create a complex living organism with a range of capabilities. All aspects of development and all aspects of subsequent behaviour ultimately depend on genomic information but the potential will not be fully realised if the environment is not conducive.

The environment can influence development in many ways. The process of development involves cell division (mitosis) in which one cell becomes billions of cells after 50 or 60 doubling divisions. Exposure to radiation, tobacco smoke or other damaging chemicals will increase the rate of mutation in mitosis and increase the rate of cell death. Diet is also likely to play a part. Certain constituents of the diet contain anti-oxidants which neutralise damaging chemicals (free radicals) produced during normal metabolism. If the diet is not sufficiently varied to include anti-oxidants then further DNA damage and mutation will occur. The significance of DNA damage during development is that the information store is gradually eroded and cells that lack genes will not necessarily behave appropriately. A single individual has billions of cells but can only be regarded as an individual because the cells communicate and act in concert. Cells send out information and receive it but they can only act appropriately if they carry the information store specifying those actions. As the information is eroded by mutation the action becomes increasingly suboptimal.

The environment provides experience influencing the development of neural networks in the brain. This is education in its broadest sense. Education in the narrower sense of formal schooling is also a key part of development and will influence performance on tests of intelligence. Children who have not been taught to read will not do themselves justice in any test regardless of their innate ability.

The concept proposed is that the genes in the zygote are the blueprint for development and behaviour but the environment also has a crucial role. If the genome is complete and free of any deletions then a facilitating environment will lead to a person who has a high level of ability in a broad range of tasks. That individual will score highly on the IQ test. Deleterious mutations in the zygote, however, will impair development, impair ability and reduce the IQ score. This hypothesis is examined in relation to the key facts of intelligence noted above.

- Performance on various components of the IQ test is positively correlated:

The hypothesis predicts that the IQ score is negatively correlated with the number of deleterious mutations in the genome. Those with a few deleterious mutations will score well on all components of the test, those with the most mutations will score badly on all components, and those with an intermediate number will have an average performance. Thus if a large number of individuals take the test some will do well on all components, some badly on all components and the results will show positive correlation. The correlation coefficients, however, will be less than 1. This is because the deleterious mutations are distributed at random in the genome and different abilities will be affected to different degrees in

different individuals. Those with a few mutations will be good at
everything, those with many will be good at nothing, but most individuals
will have an intermediate number of mutations and will be impaired in one
or two components only. The number of deleterious mutations in the
zygotes of individuals who survive to be adults (mean = Q, see chapter 1)
is approximately a Poisson distribution. This is a statistical term for a
distribution produced by a random process, it is similar in form to a normal
curve or the bell shaped curve. Thus the distribution of deleterious
mutations in the genome and the distribution of IQ scores are similar.

- IQ test results correlate positively with scholastic and academic
  success.

Since deleterious mutations impair ability they will not just affect the
components of the IQ test but also other academic or scholastic tests. Thus
the hypothesis predicts that the results will correlate positively. Academic
success requires intelligence but it also requires hard work and enthusiasm
and perhaps some luck; thus correlation with the IQ will never be perfect. It
is also important to note that the IQ test is for the general population,
picking out the top 10% from the next 10% and so on. It is much less
useful for discriminating within a 10% group. Within a group the
differences are smaller and other things, such as hard work and application,
become more significant.

- IQ test results correlate with success in life in the broadest sense
  of the term:

This is at first sight a surprising result as noted in the discussion above. It
is, however, a direct prediction of the idea that deleterious mutations impair
all aspects of ability. That which is surprising becomes obvious. Deletions
in the information store impair performance on IQ tests, in school
examinations, in College examinations but also in every thing else that
requires ability; finding a marriage partner, keeping a job, getting
promotion, fighting disease, preserving health, living to a cantankerous old
age.

Let us just stop and consider how surprising this result is. A psychologist
can enter a school and test 100 children in less than half a day on a battery
of intelligence tests. These are children she has never seen before.

Following the marking and analysis it is possible to place the children in rank order from top to bottom. That list will correlate positively with every measure of success in life that one can imagine. The list predicts the future in a way that no other psychological test, or any other scientific test, can do. When stated in this way it sounds intrinsically unlikely, but that is what the IQ test offers. If a scientific theory predicts (best of all) or explains (second best) something that is intrinsically unlikely and that intrinsically unlikely idea proves to be true then it provides strong support for the theory. Random deletions from an information store that specifies ability will impair every aspect of success that depends on ability; that is the theory and that is in practice what is found.

- There is a genetic component to intelligence:

This fits directly with the hypothesis. The distribution of intelligence (a Gaussian curve, or normal curve or bell shaped curve) is similar to the distribution of deleterious mutations in the genome. Parents with low levels of deleterious mutations will tend to have children with low levels of deleterious mutations therefore intelligence will be familial or inherited. Siblings will tend to have similar levels of deleterious mutations and therefore have similar IQs. Monozygotic twins have the same number of deleterious mutations in the germ line (the number in the zygote) and therefore will have very similar IQs. Any difference between them will be due to the environment. (The zygotes of monozygotic twins are identical and have the same number of deleterious mutations, but subsequent cells will acquire additional mutations at random during mitosis so even monozygotic twins are not genetically identical).

- There is an environmental component to intelligence:

Radiation, chemical pollutants and poor diet will increase the rate of mutation in cell division as noted above. The germ line (zygote) information store is copied in cell division to form the somatic (cells of the adult) cell information store. Any impairment of this store due to somatic mutation will impair ability and influence IQ.

We learn from experience and performance on any task depends on past experience and the ability to profit from it.

Intelligence

Achievement depends not only on ability but also on interest and application; these qualities are influenced by experience and background.

- There are racial differences in mean IQ score:

The mean IQ of Blacks in the USA is below that of Whites. In the past, those who regarded the IQ as a measure of innate ability were led to the conclusion that Blacks were inferior to Whites. What does information theory contribute to this debate?

The first point to make is that IQ tests were developed by White males; helped more recently by White females. They reflect White culture and the things that Whites consider important. We should not assume that Blacks consider them equally important and that they automatically have the same enthusiasm for doing well. I regard this as a valid but relatively weak argument. The reason I believe it is weak is that, although there are cultural and motivational differences between racial groups, we are in fact more similar than we are different.

A more substantial argument relates to the environment. In general, in the USA, Blacks have a poorer standard of living than Whites. They are more likely to be exposed to pollutants, to have a poor diet and to have a relatively poor education. All these factors will influence IQ.

The question we need to take head-on, however, is "are there genetic differences?" For a large part of history Blacks and Whites have led separate lives and have not inter-married. In the process the two groups have acquired different neutral mutations. Thus there are genetic differences related to those neutral mutations. Neutral mutations, however, are not better or worse, they are just different. It is quite possible that neutral mutations in one racial group mean that they are better at some particular ability, but it also means that they will be worse at something else, otherwise the mutation would not be neutral. Neutral mutations might improve performance on a component of the IQ but it is unlikely, indeed contrary to definition, that they could improve performance on all aspects of the IQ. In so far as Blacks and Whites are genetically different, that difference is not about being better or worse it is about being different but equal.

There is, however, another twist to the story. A good standard of living provides protection from the elements, avoidance of pollutants, a good all year round supply of nutrients, hygiene and relative freedom from infectious disease. This means a decrease in the rate of somatic mutation (mutation in cells after the zygote stage) as previously described. If there are more mutations between zygotes and gametes then P (the mean number of new mutations arising per human generation) will rise. If P rises then Q will rise. If Q rises in a racial group then the mean IQ of the group will fall, (see figures 2.2 and 2.3).

This is a crucial point and requires some clarification and expansion. In chapter one P and Q were defined. Q is the mean number of deleterious mutations in the germ line of adults i.e. the mean number of deleterious mutations in zygotes which survive to become adults. P is the mean number of new deleterious mutations that arise in cell division after the zygote stage and are passed on to the next generation. Thus cells in adults in the reproductive years have an average of P + Q deleterious mutations (Q in the zygote and P added subsequently). The zygotes that are formed in the next generation will have a mean of P + Q deleterious mutations but some will have more and some will have less. The actual distribution is a Poisson distribution with a mean of P + Q (see figure 1.4 in chapter one). The zygotes at the upper end of the distribution are less likely to survive and therefore the mean number of deleterious mutations in zygotes that survive to become adults is less than P + Q. Over several generations the population will reach equilibrium and the fall by selection will equal the rise due to new mutation. Thus a mean of P + Q in zygotes becomes a mean of Q in those zygotes that survive to become adults. In this chapter it is argued that Q is an important genetic determinant of IQ.

Let us now consider a racial group in which the mean number of new deleterious mutations per generation is Pa and the mean number of deleterious mutations in the germ line of adults is Qa. Zygotes will have a mean of Pa + Qa, and the fraction of zygotes that survive (Za) will have a mean of Qa. Let us now imagine that environmental conditions worsen and the number of new mutations per generation rises to Pb. In the next generation zygotes will have Pb + Qa mutations, the percentage of zygotes that survive will decrease (<Za) and the mean number of mutations in the zygotes that survive will be > Qa. If this is followed over several generations the percentage of zygotes that survive will gradually decrease

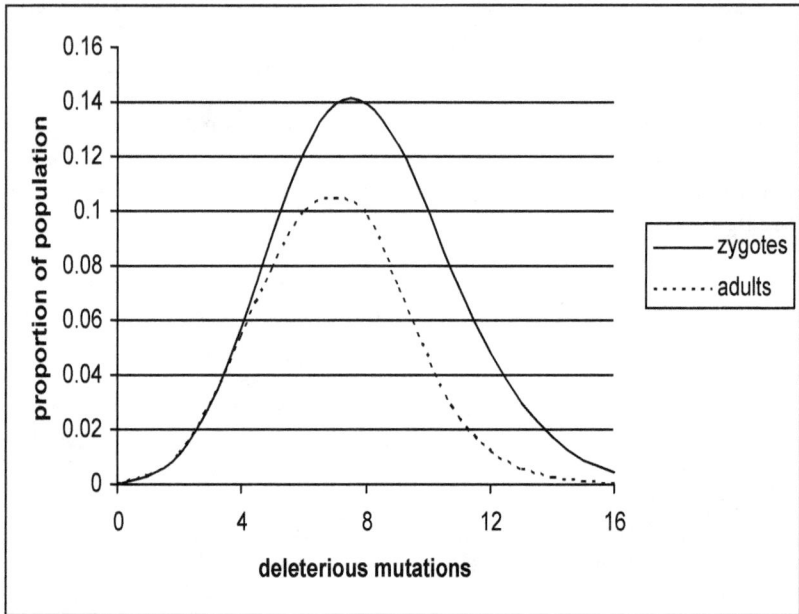

Figure 2.2

The distribution of deleterious mutations in zygotes and in adults (i.e. in zygotes that survive). In this case P=1 and Q=7.

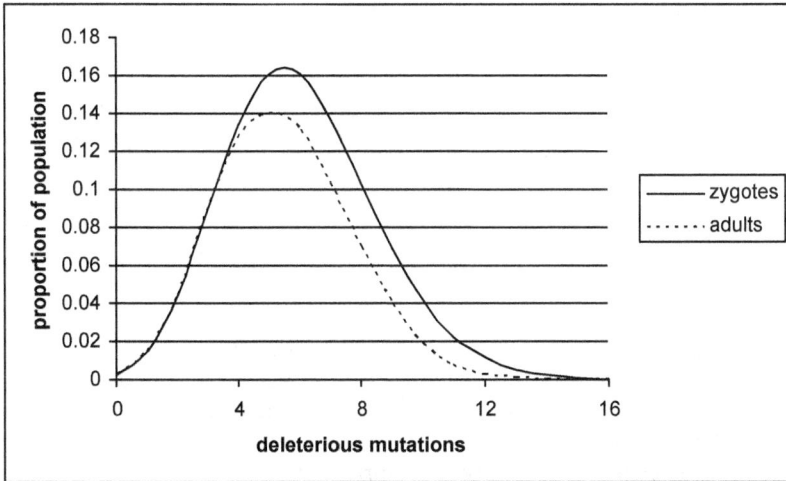

Figure 2.3

The distribution of deleterious mutations in zygotes and in adults (zygotes that survive) with P= 0.6 and Q= 5.4. Compared with figure 2.2 a fall in P leads to a fall in Q and an increase in the proportion of zygotes that survive to become adults. If P falls gradually Q will fall even more gradually over several generations before a new equilibrium is reached. It is suggested that the population of adults in figure 2.3 will achieve higher scores on IQ tests than will the population in figure 2.2 and the mean number of deleterious mutations in zygotes that survive will increase until a new equilibrium is reached. Let us designate this Pb, Qb and Zb. (Note that Pb > Pa, Qb > Qa, Zb < Za). In this example, as environmental conditions worsen and P rises then Q will rise over several generations to a new equilibrium. Equally if two racial groups experience different environmental conditions over many generations such that in one group P is lower than in the other, then the value of Q will be lower in that group and they will do better on IQ tests. If environmental conditions subsequently change so that P equalises between the two then Q will also equalise but it will take several generations. The IQ results will converge over this period.

A racial group is a collection of individuals with a common ancestry stretching over many generations. They acquire neutral mutations and deleterious mutations. There is selection against the latter but not the former. The deleterious mutations are distributed at random during sexual reproduction creating inequality in offspring. Deletions from the information store impair ability and therefore within a racial group there will be a range of ability which is genetically determined and reflects the unequal distribution of deleterious mutations between individuals. Different racial groups will have a slightly different mean number of deleterious mutations and this will lead to a different mean IQ; but as the environment improves for all racial groups the mean number of deleterious mutations will fall in all the groups and the means of the IQ will converge. There is nothing intrinsically inferior about the black genome compared with the white genome; but blacks might carry more deleterious mutations as a result of generations of social and economic disadvantage. It will correct in a few generations if social and economic advantage is maintained.

Finer Discrimination

The IQ test is a useful test of the general population. It can be used to divide up the population into five or ten groups ranked by intellectual ability. Those in the first rank will be cleverer than those in the second rank and so on down the list. But the IQ test is much less useful in discriminating within a rank. We can conclude that the top 10% are cleverer than the next 10% but we cannot conclude that the top 1% is cleverer than the next 1%.

There is a considerable range of ability within the group of individuals ranked in the top 10% of the IQ distribution; but that range of ability does not correlate strongly with the IQ score. Why should this be?

Let us assume that the mean number of deleterious mutations in the human genome is eight (Q=8). If this is the case 0.3% of the population will have either no mutations or just one mutation; 1% will have 2 mutations and 3% will have 3 mutations. The human genome is composed of a number of different systems concerned with different functions. The essence of the concept of redundancy, as discussed in chapter one, is that a single deleterious mutation in any one system has no observable or measurable effect. It is only when there are two or three or more deleterious mutations in any one system that impairment will be measurable. Thus IQ tests cannot discriminate within the top 0.3% of the population as there is no difference between no mutations and one mutation. Furthermore they are unlikely to be useful in discriminating within the top 1.3%; because in individuals with 2 mutations it is unlikely that the mutations will be in the same system therefore there will be no difference between no mutations, one mutation and two mutations. Even discrimination within the top 4.3% will be limited to those few individuals who have 2 mutations in one system; the rest will be indistinguishable.

The IQ test divides the general population into roughly ten groups, with meaningful distinction between the groups but no meaningful distinction within the groups. This fits very well with the idea that the difference between groups is determined by the number of deleterious mutations in the genome. There obviously are differences in ability between individuals within the top 10% of the population but these are not usefully measured by the IQ and are not determined by deleterious mutations. Differences within groups depend on neutral mutations and environmental experience.

Temporal Changes in Population IQ

Measured IQ has risen by 15 points since the tests were first introduced. This is most probably a consequence of improved education in its widest sense. But the environment can also improve IQ in other ways. Improved socio-economic conditions can act to reduce the rate of somatic mutation in cells and thereby reduce P (the mean number of additional new mutations per generation). If P falls then Q will also fall as indicated above. A fall in Q will lead to an increase in mean IQ.

If the mean number of genomic deleterious mutations falls from 8 to 6, which is feasible over several generations, then the mean population IQ will rise by over 10 points. Thus, although mediated by the environment, the rise in IQ could be due to genetic mechanisms. This underlines the importance of thinking of nature and nurture not nature or nurture.

It is commonly stated that the rise in IQ cannot be due to genetic changes because they occur over a much longer time scale. This is true of conventional genetics; environmental selection for or against a specific gene takes many generations to produce a measurable change. It is not true, however, of the mechanisms discussed in this book.

Conclusion

Biological organisms are made up of a number of complex highly redundant systems specified by genes. All aspects of performance are determined by an interaction between the environment and the systems as specified. Deleterious mutations (heterozygote deletions) interact synergistically to impair performance. Measures of intelligence (IQ) are just one aspect of the overall performance of an organism; the measures will correlate negatively with the number of heterozygous deletions in the genome.

Chapter Three

Infection

Introduction

Shortly after birth we are colonized by a bacterial flora. Bacterial organisms are found on the skin, on the surface of the upper respiratory tract and in the gut. The vast majority of these bacteria are harmless commensals. They grow on the surface but do not invade the body and do not cause disease.

A few bacteria, however, are pathogenic i.e. they are capable of causing disease. These organisms grow on the body surface along with commensal bacteria but cause damage either by the secretion of bacterial toxins (toxic chemical products secreted by bacteria) or by invading the body.

We have many defences against invading bacteria. In fact our defences form a highly complex and a highly redundant system. In the gastrointestinal tract for instance, the surface cells secrete a mucus layer which forms a physical barrier to invasion. If the bacteria breach the barrier then inflammation ensues and neutrophils move from the bloodstream into the tissue and phagocytose and destroy the bacteria. The process of phagocytosis is one in which the neutrophils engulf bacteria and then release bags of poison onto the bacterial surface and cause their rapid death. Some pathogenic bacteria, however, have a mucus layer on their surface which impedes phagocytosis. Neutrophils then need help from molecules which coat the bacterial surface and aid phagocytosis. Some of these molecules are part of our innate defences; some are part of a system of acquired immunity.

Acquired immunity depends on lymphocytes of which there are two main types – T lymphocytes and B lymphocytes. B lymphocyes produce immunoglobulins which can be present on the surface of the lymphocyte or can be secreted into blood or onto the body surface. Immunoglobulin molecules specifically recognize and adhere to chemical groupings (epitopes) present on the surface of bacteria. One end of the immunoglobulin molecule has a shape complementary to the epitope and sticks to it; the other end of the molecule can then adhere to the surface of a phagocytic cell and aid bacterial destruction. The mechanism is complex in that the immunoglobulin adherent to bacteria will trigger changes in blood proteins called complement which coat the bacteria and further aid phagocytosis. B lymphocytes can mature into plasma cells which secrete large amounts of immunoglobulin. There are several types of immunoglobulin called IgA, IgM, IgG, IgD and IgE. IgG is the main immunoglobulin found in the blood. IgA occurs on body surfaces such as the gastrointestinal tract. IgE occurs on the surface of mast cells which are important in asthma (see later).

T lymphocytes are concerned with the distinction between self and not self. Bacteria in the gut are phagocytosed by macrophages (big phagocytes rather than neutrophils which are little phagocytes). The macrophages kill the bacteria and then break their many proteins into short sequences of 8 to 12 amino acids which are posted onto the surface of the cell in association with a HLA surface protein. The T lymphocyte has a specific receptor that recognizes a particular amino acid sequence in association with the HLA protein. In ways which are not fully understood the T lymphocyte must recognize sequences from pathogens and initiate a B lymphocyte response to generate immunoglobulin which recognizes foreign, not self components.

If a specific pathogenic bacterium is to cause significant disease then it is most likely to occur on first exposure, because there is only innate immunity and no acquired immunity. Subsequent exposure will quickly activate acquired immunity and the infection will be less likely to develop.

Bacteria are unicellular organisms that can grow outside the body. They have DNA which is their information code; they also have RNA and manufacture proteins and other cell components. Viruses are naked DNA or RNA with a protein outer coat but they cannot grow outside cells. Thus viruses cause intracellular infections. In an upper respiratory tract infection, for instance, viruses are spread from person to person by coughs and sneezes. The virus particle adheres to the surface of a respiratory epithelial cell and then enters the cell. The protein coat dissolves and the viral DNA or RNA uses host intracellular enzymes to replicate and convert one strand of DNA or RNA into thousands of strands. Each strand is then coated by protein courtesy of the host. The epithelial cell then dies and releases the viral particles to infect other cells in the same person or a different person. Acquired immunity depends on T and B lymphocytes which specifically recognize the viral protein coat. Subsequently IgA in nasal secretions will neutralize the viral protein and prevent the virus entering the cells thereby preventing further infections.

The above is a brief outline of how bacteria and viruses can cause disease and how the body responds. The system, however, is very complex and far from fully understood. A complex system, however, can be analyzed in terms of information theory without necessarily knowing all the details.

If a potential pathogen is present on the gut surface or on the respiratory tract surface, the task of the immune system is to recognize epitopes that resemble self and distinguish them from epitopes that do not resemble self (foreign epitopes). Clones of T and B lymphocytes will be committed to attacking foreign epitopes in order to prevent infection. But it is essential that epitopes resembling self are not attacked or there is a risk of autoimmune disease. Equally it is important that a distinction is made between potential pathogens and commensals. If clones of lymphocytes are committed to attacking commensal organisms then there is a risk of inflammation and tissue damage for no advantage.

The size of the problem faced by the immune system in responding to the bacterial flora is immense:

Humans have 25,000 genes in the haploid set (50,000 in the diploid set). These genes code for in excess of 250,000 proteins. Each protein has on average 300 amino acids and very many individual epitopes. These are all potential self epitopes which could be involved in autoimmune disease if a mistake occurs. Bacteria have perhaps 5000 genes coding for 5000 proteins. But there are at least 400 species of bacteria that can reside in the gut and there will be approximately 40 in any one individual at any one time. Thus the proteome of the bacterial flora is of the same order of magnitude as the proteome of the human. Distinguishing self from non self is made more difficult by the fact that genes are highly conserved in evolution. DNA repair enzymes, transmembrane ionic channels, and many protein enzymes and secretory products evolved in bacteria and have been conserved in evolution. It is therefore no exaggeration to claim that we share many of our genes with the rest of creation and there is therefore considerable potential overlap between our proteome and that of bacteria.

The T lymphocyte recognition system is a surface receptor that recognizes an 8 to 12 amino acid sequence sitting in the groove of an HLA molecule.

There are 20 amino acids and the number of ways they can be arranged in a sequence of 10 is 20^{10} which equals 10^{13} (i.e. ten thousand billion). There are approximately 10^{13} lymphocytes in the body of an adult but since each clone must have hundreds if not thousands of cells the number of clones is only a fraction of this number. One of the unsolved mysteries of immunology is how T lymphocytes learn to ignore the majority of amino acid sequences in the HLA groove.

The basic laws of complex information processing systems set out in chapter one include the statement that there is always a finite chance of error when information is analyzed in uncertainty. The errors that can occur in responding to the microbial flora are:

- Type one: Failure to recognize an epitope on a pathogen as foreign. This will increase the risk of infection.
- Type two: Failure to recognize an epitope on a pathogen as similar to self. This will increase the risk of autoimmune disease.
- Type three: Failure to ignore foreign epitopes on commensal bacteria. This will lead to inflammation in the respiratory tract and gut in response to commensal bacteria. Asthma in children is commonly provoked by exposure to faecal antigens of the house dust mite. Could this be due to a failure to ignore that which should have been ignored?

First Exposure to Microbes

Most pathogenic bacteria and viruses are relatively common. We tend to meet them early in life and the diseases they cause are most commonly found in childhood. Figure 3.1 shows how the probability of first exposure to common organisms changes with age. If 80% of the population meet an organism in any one year then 80% will be exposed for the first time in year one, 80% of the remaining 20% will be exposed for the first time in year two and so on. The age incidence of first exposure will fall rapidly with age. If only 10% are exposed in any one year then 10% will be exposed for the first time in year one, then 10% of the remaining 90% will be exposed for the first time in year two and so on. In this case the age incidence curve will fall more slowly. In general the age incidence of first

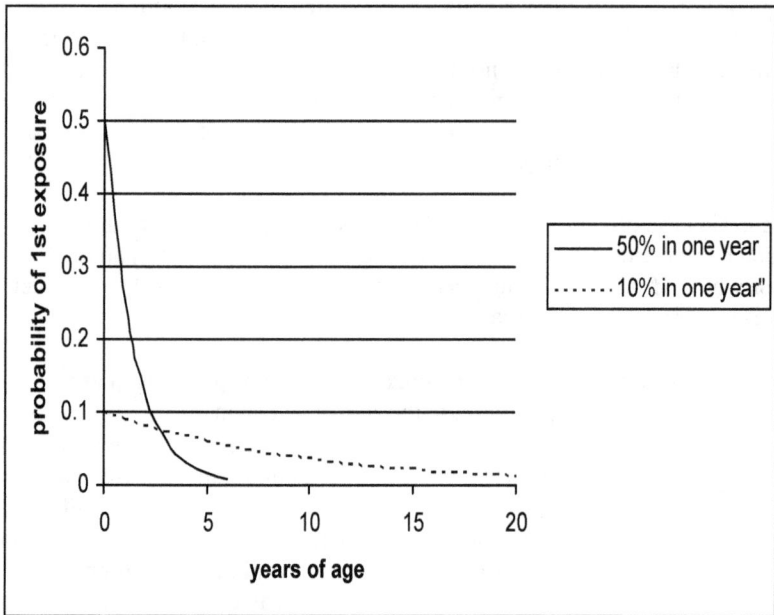

Figure 3.1

The probability of first exposure by age for an organism that 50% of the population meet in any one year and for a less common organism that 10% of the population meet in any one year. Disease on first exposure is unlikely for the more common organism after 5 years of age but can still occur for the less common organism in the teenage years.

exposure will be Ce^{-kt} (where C is a constant, e the exponential function, k is a constant specifying the rate of exposure and t is time.)

The probability of making a mistake on first exposure will, however, rise with time. This depends on the fundamental laws of information theory. Consider a complex system responding in an uncertain world. Let us define the probability of error as $1 - R$, where R is the probability of a correct response. This is the second law i.e. there is always a finite chance of error in a complex information processing system. The third law states that systems decay at random with time. This means the probability of error should be written as $1 - Re^{-pt}$ (where e is the exponential function as before, t is time and p is the rate constant for decay with time). The fourth law states that complex systems need to be highly redundant. This can be modelled by considering not one decision system, but n systems working in parallel. They only get it wrong if all the systems make a mistake at the same time. The probability of error is then $(1 - Re^{-pt})^n$. A mistake in a complex system will lead to disease and the ultimate mistake will lead to death. This error function rising with age is plotted together with the mortality rates in figure 3.2. The function predicts the overall form of the mortality curve.

In chapter one the point was made that the question to ask concerning a prediction of a mathematical model is not "is it correct?", but "is it useful?" Mathematical theories and models cannot be correct because they are always simplifications of a more complex world. In the last analysis they are always wrong. The key question is, "is the model useful?" If the model preserves the essence of the problem then it will be useful. The above simple function cannot be a complete theory of ageing. But the essence of ageing is a complex highly redundant system decaying at random with time; the function has those key features and thereby preserves the essence of ageing. At the very least it matches the age incidence curve of mortality and we should stick with it.

Disease on first exposure to microbes therefore has two components. The probability of meeting an organism for the first time falls with time, but the probability of an error on first exposure rises with time. The resultant risk of disease is obtained by multiplying the two curves. The product is an age incidence of disease which rises to a peak and then falls. If the organism is common the disease will peak early in life; if the organism is less common

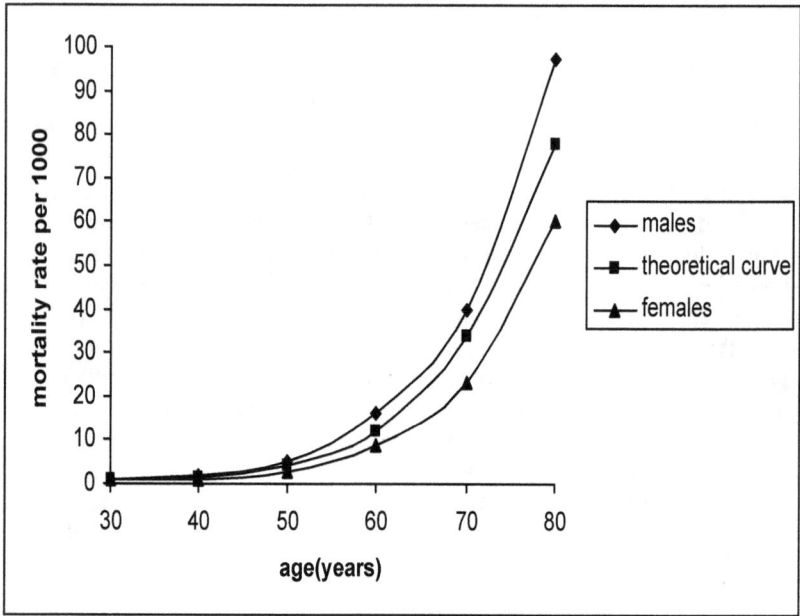

Figure 3.2

The mortality rate by age for men and women in England and Wales in 1988 together with a theoretical curve generated by the function $(1 - Re^{-pt})^n$

.

the peak will occur later in life. Furthermore the less common the organism the later the peak and the higher the peak. In other words if organisms circulate more slowly leading to later exposure the frequency of the disease will increase.

Multiple Sclerosis

Multiple sclerosis is an autoimmune disease in which lymphocytes attack and damage myelin sheaths which surround and protect nerve fibres in the brain and spinal cord. The age incidence of the disease onset is shown in figure 3.3. The risk of disease rises to a peak in the late twenties and early thirties. The chance of a first attack then steadily falls. The hypothesis is that first exposure to a particular micro-organism (it could be viral or bacterial) leads to an autoimmune attack on a microbial epitope which is similar in structure to an epitope present on myelin. The age incidence curve fits with an organism that 50% of the population meet in a four year period.

The disease shows a pattern of remission and then exacerbation followed by a further period of remission. This fits with recurrent exposure to micro-organisms. First exposure to the organism leads to an autoimmune attack on myelin and symptoms occur due to nerve damage. The organism is then eliminated and the nerve damage heals with perhaps some residual disability. Further exposure to the organism, perhaps four years later, leads to a further attack of multiple sclerosis with further nerve damage and another set of symptoms. The organism is eliminated once more and the new damage heals but once again with some residual disability. It is important to realize that we are continually re-exposed to common viruses and bacteria. An infection that leads to a clinical illness is most likely to occur on first exposure, on subsequent exposure we have an immunological memory of the organism and quickly produce a specific attack on the organism. But if the organism has triggered an autoimmune attack on a body tissue then exacerbations of autoimmunity will occur with each episode of re-exposure.

A feature of multiple sclerosis, which it shares with many other autoimmune conditions, is that the myelin damage with each exacerbation

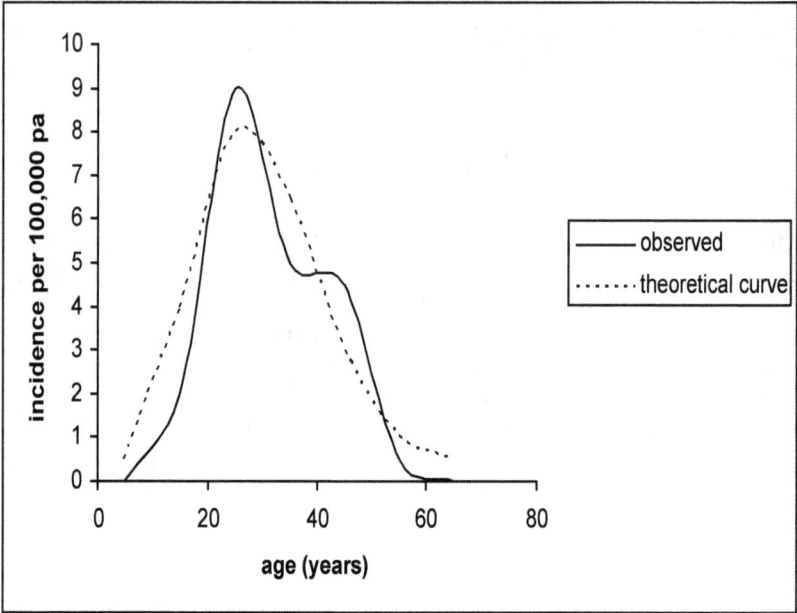

Figure 3.3

The age incidence of multiple sclerosis from observation plotted together with a theoretical curve derived by multiplying Ce^{-kt} by $(1 - Re^{-pt})^n$. The theoretical curve fits with an organism that 50% of the population meet in approximately 4 years.

of disease is localized. The explanation for this is unclear; all myelin presumably shares the problematic epitope so it is a challenge to try to explain why just a small focus is damaged at any one time.

Another feature of multiple sclerosis which it shares with other autoimmune conditions is that HLA antigens have a role in causation. The HLA molecule sits on the surface of human cells and T lymphocytes recognize short polypeptide molecules sitting within a groove in the HLA molecule. The mistake that leads to multiple sclerosis occurs at this point of recognition. During evolution neutral base changes have occurred in the genes that specify the HLA molecule so that there are a number of different types. It appears that certain types of HLA increase the risk of multiple sclerosis. Presumably they decrease the risk of other diseases and therefore are neutral overall. One imagines that when the polypeptide related to myelin sits in the HLA molecule it might be partially obscured and more difficult to read and recognize depending on the precise structural configuration of the HLA molecule. This is an example of how a neutral genetic change can increase the risk of a particular disease.

Multiple sclerosis becomes more common with increased distance from the equator, it is more common in colder climates and it is a disease of technologically advanced nations. If children grow up in a low incidence area and then emigrate as adults to a high incidence area their risk is low. If they grow up in a high incidence area and then emigrate as adults they have a high risk. (In fact the term high risk is somewhat misleading because even in so called high incidence areas the vast majority of the population do not get the disease.) This can be explained if we remember that bacteria and viruses spread more slowly when hygienic standards are high and when the climate is cold. Slowing the rate of spread of the organism will mean that more individuals meet the organism for the first time as adults and therefore the incidence of the disease will rise. Thus the disease is increased in advanced nations with cold climates; those who spend their childhood in these areas might not meet the organisms and will be at risk as adults. Those who grow up where the organisms spread more rapidly are likely to meet the organisms as children and are unlikely to get the disease.

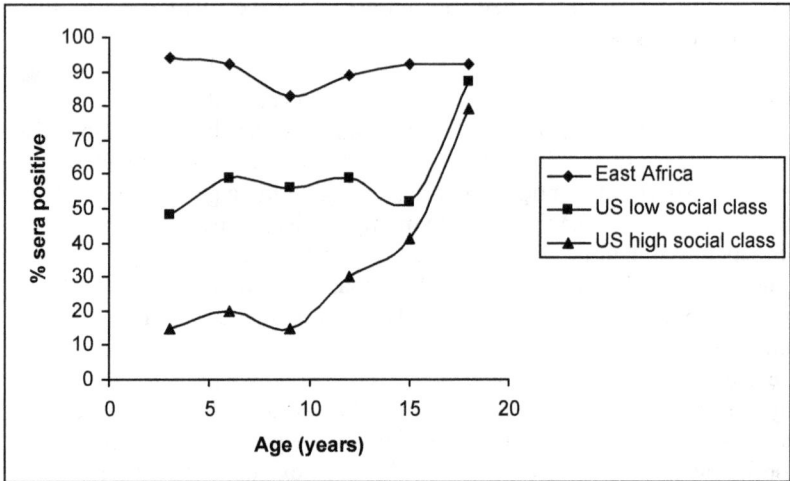

Figure 3.4

Age related incidence of positive serology for antibodies specific for Epstein Barr virus. The graph indicates that children in East Africa meet the virus early in life, US high social class children meet it later and US low social class are intermediate.

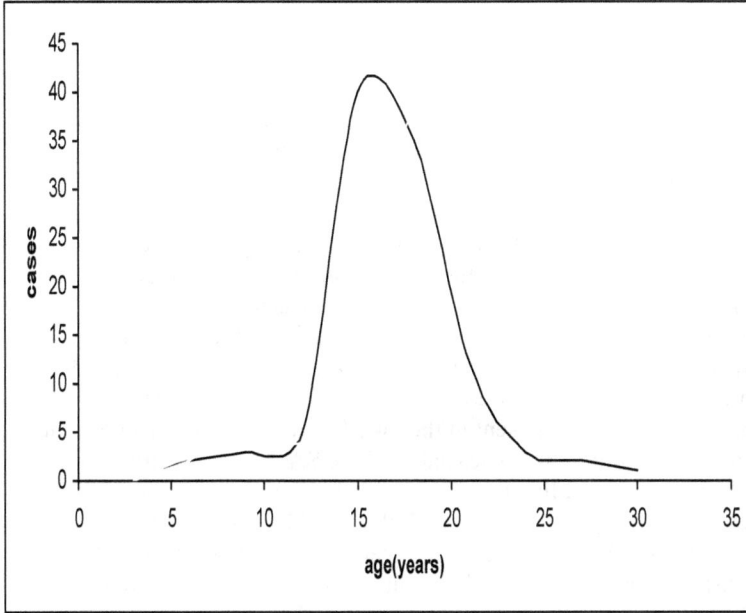

Figure 3.5

Cases of glandular fever from the USA recorded in 1968

Glandular Fever

Glandular fever is a disease then tends to strike teenagers. It is caused by a virus called the Epstein – Barr virus. It is spread by the intimate sharing of body secretions as in kissing and other forms of close contact. The virus enters the lining cells of the throat and then spreads to infect lymphocytes. Viral infection of lymphocytes causes cell proliferation and as a result lymph glands enlarge. The victim feels ill, has a sore throat and enlarged tender lymph nodes in the neck. Abnormal lymphocytes appear in the blood and this appearance together with other changes in blood proteins allows the diagnosis to be made. The disease lasts a few weeks and then slowly abates as the immune system reduces the viral load, but the organism is often carried for life. It is possible to test for previous infection by Epstein – Barr virus. Those who have met the virus have specific anti-viral antibody which is present in the bloodstream. Figure 3.4 shows the prevalence of antibody in Africa and in the USA. In Africa all children meet the virus early in life, they have only a minor infection and they do not get the disease glandular fever. Children in the lower socio-economic classes in the USA are also likely to meet the virus in childhood; they have a low risk of glandular fever. The group with the highest risk are the higher social class groups in the USA. They are less likely to meet the virus in childhood and therefore are at risk as teenagers. The virus causing glandular fever is more common than the organism that triggers multiple sclerosis, because the figure shows that the vast majority has met the organism by the end of the teenage years. Therefore the model developed above predicts that the disease glandular fever will peak earlier than multiple sclerosis and that is what is found. (Figure 3.5). Multiple sclerosis peaks in the late twenties and early thirties, glandular fever peaks in the late teenage years.

Type One Diabetes Mellitus (Insulin Dependent Diabetes Mellitus)

Type one diabetes mellitus is also an autoimmune disease. In this condition lymphocytes attack and destroy the cells in the pancreas that manufacture insulin. The attacks occur in waves followed by periods of inactivity.

Eventually, after multiple waves of attack, the majority of the insulin producing cells are destroyed and the child or young adult presents with a clinical disease. Often there is an intercurrent viral infection such as a common cold but the child then becomes very thirsty, passes increased amounts of urine and the blood glucose rises. Without treatment death will follow. The treatment is to give insulin. A so called honeymoon period often follows the first attack. For a while the child returns to good health and does not need insulin. The reason is that the last wave of destruction in the pancreas subsides and a few insulin secreting cells recover. Furthermore the demand for insulin is less when the child is free of the intercurrent infection. But then another wave occurs, all the insulin producing cells are destroyed and the need for insulin is life long.

Clinical presentation with type 1 diabetes mellitus peaks at around age 11 or 12. This peak is a lot earlier than for multiple sclerosis and earlier than glandular fever. According to the model we have developed this implies that the causative organisms that have triggered the disease must be much more common. But in fact the model is not actually concerned with age of presentation of diabetes which is late in the course of the disease but with the age of the first wave of attack on the pancreatic cells. The peak age for first attack must be less than 12 and is probably in the first few years of life. This means that the causative organisms must be very common indeed. There are very few viruses that most of us meet in the first one or two years of life, but there are many bacteria which we all meet in the first few years of life. The bacteria which colonise our upper respiratory tracts and our gastro-intestinal tracts are very common, we meet them early and if they trigger an autoimmune disease it will start early.

Thus the model leads to an interesting prediction. It is that bacterial infection triggers type one diabetes mellitus. This prediction is directly contrary to conventional wisdom, but it demonstrates the power of deduction from first principles. The prediction may or may not be true, only time will tell. But in the meantime we should search for evidence of molecular mimicry between common organisms of our microbial flora and pancreatic insulin producing cells.

Type one diabetes mellitus is more common in first born children than in subsequent siblings. It is more common in those of higher social class. There is some evidence that breast feeding protects. Early exposure to other children by attending nursery in the first year of life also protects. All these features fit with the idea that early exposure to micro-organisms is protective.

HLA molecules also have a role in pathogenesis. Diabetes is more common in association with certain HLA types; we must presume that the combination of the HLA molecule and specific polypeptides relevant to diabetes is in some way more difficult to read.

Asthma, Eczema and Hay Fever (Atopic Disease)

There has been a marked increase in the frequency of these conditions over the last 30 or 40 years. Childhood asthma used to be rare and is now common in the western world. These conditions are associated with improved living standards, with technological advance, with better socio-economic conditions, with small sibship size. They are more often found in first born than subsequent siblings. Rural isolation increases the risk but exposure to farm animals lowers the risk.

The house dust mite feeds on exfoliated skin cells and is commonly found living on pillows and blankets. Inhalation of faeces from the house dust mite can precipitate asthma. Antigens (chemicals) present in faeces combine with IgE molecules sitting on the surface of mast cells. The combination of antigen and IgE triggers the release of inflammatory chemicals stored in the mast cell. The released chemicals cause inflammation and there is increased secretion of mucus in the airways and swelling of the wall of the airways. This causes wheezing, which is the characteristic feature of asthma. In the case of hay fever, pollen antigens combine with IgE on mast cells in the nasal cavity and cause release of chemicals and inflammation. This produces a runny nose and sneezing etc. Eczema is commonly found in association with asthma and hay fever although the mechanism is somewhat different and less well understood.

One of the tasks of the immune system is to learn to ignore the epitopes that should be ignored such as those on commensal bacteria. An error in this process is labelled a type three error above. It is tempting to suggest that atopic disease is a consequence of type three error. The faecal flora of the house dust mite will have many epitopes in common with human faecal flora. If type three error has led to immunoglobulins that recognize commensal epitopes then a reaction to house dust mite can be understood.

Gene – Environment Interactions in Disease

Genetic and environmental factors influence the risk of disease. The immune response to micro-organisms is established at the time of first exposure. Errors can increase the risk of infection, of autoimmunity and of atopic disease. Information theory tells us that errors cannot be eliminated. But there are factors that can increase and decrease the risk.

The dose of the organism and the virulence factors it carries will influence risk. The age of first exposure is also a crucial factor as argued above. Factors that influence the general health of the host, such as diet, will also be important.

Genetic constitution also plays a part and we must consider both neutral and deleterious mutations.

Neutral mutations occur during cell division and are passed onto the next generation. There is no selection against neutral mutations and they gradually accumulate in the genome. The result is that our DNA differs at approximately 1 in every 300 bases. The majority of these changes are in bases that do not code for proteins. But some base changes do occur in protein coding sequences. The result is a protein whose amino acid sequence is altered. A neutral change means that the altered protein is just as successful as the original protein. It might differ in action but the difference is neither better nor worse. Or more precisely we should say there is no measurable difference.

Neutral mutations can, however, be advantageous when rare. For instance consider a protein on the surface of an epithelial cell in the upper respiratory tract. If the protein is common then it is likely that there will be pathogens that have evolved to adhere to the protein in order to initiate infection. An individual with a slightly modified protein will then be at an advantage. That person will be less likely to succumb to the infection. As a result the rare neutral mutation will gradually increase in number and become common. The consequence, however, is that pathogens will evolve that can adhere to this mutated protein and its advantage will be lost.

A related idea is that of heterozygous advantage. Consider an individual with different genes at a particular locus rather then the same gene. If the locus codes for an epithelial surface protein the heterozygous condition could be advantageous. Homozygous individuals would be at risk of heavy colonization because all the surface molecules would be the same. But heterozygous individuals would have half the surface molecules of one type and half of another. Colonization by any particular pathogen would be reduced to one half.

Neutral mutations can also influence regulatory elements that control the expression of genes related to infection. For instance the inflammatory response to infection is controlled by a complex network of cytokines (molecular messengers). Expression of pro and anti inflammatory cytokines is controlled by regulatory genes. Thus neutral changes in the genome can lead to an increased or a decreased strength of response. The change is neutral becomes in some situations an increased response will be advantageous and in other situations a decreased response will be advantageous. But what if an individual is heterozygous for the mutation? That individual will have two strategies at their disposal; if deployed appropriately this will give the heterozygote an advantage over the homozygote.

If we think carefully about neutral mutations then we only know a mutation is neutral if it survives in the genome but does not displace the original gene. This is most likely to occur if the mutation is advantageous when rare but neutral when common. Furthermore it is most likely to survive for a long period of time if there is heterozygous advantage because this prevents any one of the two forms of the gene displacing the other. All this is most easily understood if the genes under consideration are related to infection.

This leads to an important consideration concerning the difference between boys and girls. The X chromosome has 5% of the genome. Females have two X chromosomes and males have only one. Males do have a Y chromosome that females lack; but the Y is a squidgy little thing and does not amount to much in information coding terms. Females inactivate the X chromosome at random in their cells so that one half have one X and the other half the other X. This means that only females have heterozygous advantage on the X chromosome. Males enjoy heterozygous advantage for 95% of loci but females have heterozygous advantage for 100% of loci. This gives females a great advantage in fighting infection.

But there is an interesting corollary to this point. In information terms, or in terms of decision theory, infection is a sin of omission; a failure to respond to that which should be responded to. Females have two sets of cells (different X chromosomes) and both set would have to make a mistake for infection to occur. Thus females are at lower risk of infection. But autoimmune disease is a sin of commission; a response to that which should not be responded to. Females have two chances of making that mistake, males only one. Thus females are at increased risk of autoimmune disease.

Deleterious mutations are also liable to play a part in the risk of infection. We all carry a number of deleterious mutations in our cells. The average zygote that survives to form an individual has approximately 6 to 10 deleterious mutations. They act synergistically to impair performance in all aspects of biology. The lucky few with a small number of deleterious mutations will have systems that function well and the number of errors in information processing will be small. They will be relatively protected against infection. Those at the other end of the spectrum with a high number of deleterious mutations will be at increased risk of errors in decision making and therefore at increased risk of both infection and autoimmune disease.

Second and Subsequent Response

First exposure to an organism leads to an acquired immune response and the memory of that response is life long. The risk of infection and the risk of triggering autoimmune disease are most likely to occur at this time. But disease can occur on second or subsequent exposure.

Infection

Viral infections such as measles, mumps, rubella etc. cause disease on first exposure. If the organism is successfully eliminated immunity is life long and a second infection is unlikely. First exposure to Epstein-Barr virus causes no apparent disease or glandular fever. Following exposure the virus can be carried within our cells for the rest of our lives but the disease glandular fever does not recur.

Bacteria such as *Staphylococcus aureus* are met early in life. They commonly grow on the mucosal surface of the upper respiratory tract. Infants then produce an immune response to the organism and it is commonly eliminated but only to be replaced by another strain of the same organism. This is then subsequently eliminated and a further strain emerges. Infants are at risk of invasive disease from these bacteria and this could be one cause of cot death. But by the end of the first year of life infants have developed immune responses to the majority of common bacteria and the risk of cot death is vanishingly rare (see below).

In the colon a bacterial flora is established in the first year of life. There are of the order of 40 different bacterial species in each individual. These organisms are in a stable balance and it is very difficult for another organism to become established. Furthermore if a potentially pathogenic organism enters the colon it will be kept in check by the established flora. This gives the immune system plenty of time to assess the new organism and develop an appropriate immune response.

Bacteria are present on the skin, upper respiratory tract and colon throughout life. They are kept in check by a variety of means including the physical barriers provided by the surface epithelial cells and our immune systems. But these defences can breakdown. For instance burns to the skin will allow bacteria to grow and can lead to a fatal infection. If the bowel is perforated by trauma bacteria will pass into the peritoneal cavity (the space between the loops of bowel and the abdominal wall) and cause peritonitis which can be fatal. In elderly bedridden people a failure to adequately clear secretions from the lung can lead to bacterial growth causing bronchopneumonia.

The chance of these various diseases increases in frequency with age. In fact any disease that interferes with our defences can lead to infection which will exacerbate the underlying disease. Thus in terms of our information model the risk of disease on first exposure rises then falls with age; but the risk of disease on subsequent exposure rises inexorably with age. Bacteria are ever present so that the risk from the bacteria is constant, but the probability of breakdown of one or more defences is given by the ageing function $(1 - Re^{-pt})^n$.

Preventing disease

We should seek to ensure that exposure to bacteria is low dose, early and mucosal.

- *Low dose:* This gives the immune system time to analyze the bacteria and will reduce the probability of error.
- *Early:* if exposure is early in life the chance of error is reduced as described above.
- *Mucosal:* Information processing capacity is concentrated on the mucosal surface, particularly of the gastro-intestinal tract. This also raises the probability of a correct response and decreases the chance of an error.

But increasingly in technologically advanced cultures first exposure is high dose, late and parenteral.

- *Late:* Increased hygiene, small families, large houses, hot water etc all lead to decreased rate of circulation of organisms and the mean age of first exposure is increased.
- *High dose:* Late exposure to organisms increases epidemic as opposed to endemic spread. The latter is low dose and continuous, the former is intermittent, high dose and explosive. Epidemic spread means higher dose and increased risk of error.
- *Parenteral:* Increasingly first exposure is by injection of antigens in order to develop immunity (immunization). This is accompanied by adjuvants which fool the body into responding to all the antigens (remember an important part of the immune response is ignoring that which should be ignored). Mucosal exposure causes an IgA response. The IgA is secreted onto the mucosal surface and bacteria are removed without causing

damage. Parenteral injection leads to an IgG response, this removes bacteria by phagocytosis in the body and in the process inflammation ensues. Asthma is a condition in which there is an inflammatory response to epitopes that should have been ignored. I know of no evidence that immunization in childhood increases the risk of asthma, but I suspect it does. Time will tell.

Is it possible for society to change direction and achieve low dose, early mucosal exposure? The answer is a cautious yes. We need to develop methods of immunization based on mucosal exposure rather than parenteral injection. We need to engineer early low dose exposure to bacteria to build slowly a stable bacterial flora. But first of all we need to determine exactly what constitutes the normal flora and work out which bacteria are "friendly" and which are not. This is a major research programme and I believe it will be one of the big areas of scientific research over the next 50 years. The technology to carry out the research has been developed. We can sequence the genome of bacteria and determine the structure of the proteins for which the genes code (the proteome). By comparing the bacterial proteome with our own proteome it will be possible to recognize key epitopes in infection and autoimmunity.

Does Society Need a Colon?

The human colon is a remarkable structure. It is large and contains a vast number of bacteria. There are more bacterial cells in the human colon than there are cells in the human body. There are at least 400 different species of bacteria that can inhabit the colon. Each individual carries approximately 40 of these. Once the bacterial flora is established in any one individual it is stable thereafter.

One of the difficulties in studying the microbial flora of the colon is that only a minority of the bacterial species can be grown by conventional means. But it is possible using rather more expensive methods to examine the DNA content of the gut flora and deduce how many species are present. The expense means that only a limited number of studies have been carried out so far but this is an increasing area of research endeavour.

Each individual immune system carefully analyzes the bacterial flora and can influence its composition by secreting IgA molecules. The immunoglobulin can combine with and remove unwanted bacteria. In this way the flora increasingly is made up of commensal bacteria that can live in the colon without causing harm. The bacteria are not harmless because in certain situations they can cause disease. For instance the organism *Clostridium perfringens* is commonly found in the colon; if this organism gets into wounds, such as occurred in the First World War, then it rapidly causes gas gangrene and death.

Thus the colonic flora of individuals consists of selected bacteria; those least likely to lead to disease and most likely to keep potential pathogens in check. The flora is immense in terms of number of bacteria and diversity of bacteria. Each day a large number of these bacteria are voided and therefore the organisms circulating between individuals are ones least likely to cause disease. Thus a large colon with a large residue of faeces protects individuals but also protects society in general.

The safe disposal of sewage interferes with this process and, although nobody would wish to reintroduce open stinking sewers in our cities, it does appear possible that we have lost something in the pursuit of progress. Perhaps society needs an artificial colon. Large vats in which "safe" bacteria are grown and then presented in dilute form for human consumption. Maybe that is how early low dose exposure to bacteria can be engineered.

Sudden Unexpected Death in Infancy

Sudden unexpected death in infancy (SUDI) is simply defined as the death of an infant occurring between 7 days and 365 days of age which is sudden and which is not expected. If the cause of death remains unexplained after a detailed post mortem examination the death is classified as sudden infant death syndrome (SIDS). If a cause of death is found at post mortem the case is explained SUDI, if a cause of death is not found it is unexplained SUDI which is synonymous with SIDS.

Infection

The majority of cases of SUDI remain unexplained after post mortem. It is thought that the majority of these cases, classified as SIDS, are due to natural disease but we do not know the nature of that disease.

The single most consistent and characteristic feature of SIDS is the age incidence (figure3.6). In every major published series the risk of death rises from birth to peak at 8 to 10 weeks of age. The risk then falls so that the disease is uncommon after 6 months of age and falls to a very low level at 12 months of age. The definition of SIDS precludes the diagnosis after 12 months but obviously biology does not obey arbitrary rules and occasional cases of sudden death do occur after 12 months.

For the student of information theory there is a very obvious explanation for this age incidence curve. Mothers pump IgG molecules across the placenta and the infant is born with adult levels of circulating IgG. These molecules protect against extra-cellular infection by bacteria and bacterial toxins. In the first few weeks of life this maternal IgG is quickly consumed by exposure to organisms and the levels fall to a nadir at 8 to 10 weeks of age. The level of IgG then slowly rises as the infants meet and acquire immunity to common organisms. By 12 months of age infants have sufficient protection and death due to infection is rare.

If infection is a cause of SIDS the causative organisms must be very common because all infants must meet the organisms and acquire immunity in the first year of life; by 12 months of age the infants who have not succumbed to SIDS are immune. The only organisms that are sufficiently common are bacteria of the normal body flora; bacteria such as *Staphylococcus aureus* or *Escherichia coli*. These organisms are met in the first year of life and become temporary residents in the upper respiratory tract or in the gut. Viruses are less common and even in the third world there are no viruses that all infants meet in the first year of life.

The prediction is, therefore, that SIDS is caused by bacteria or bacterial toxins produced by common organisms which are found as part of our normal bacterial flora. There is now a considerable body of evidence that supports this proposition. In the 1980s SIDS was common with up to 1500 death per annum in England and Wales. At that time death was more common in the winter months when viral respiratory infections were prevalent. The concept proposed is that the viral infection led to bacterial

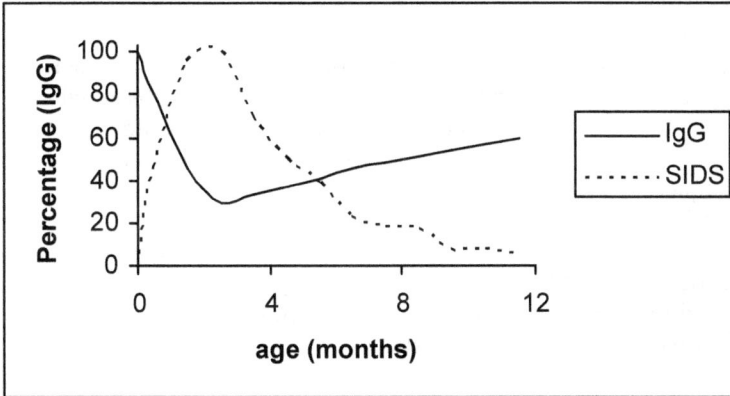

Figure 3.6

The number of cases of SIDS rises to a peak at 2 to 3 months of age. Infant IgG concentration in blood is at its lowest level at 2 to 3 months.
Information to protect against bacteria and bacterial toxins is reciprocal to the risk of death from SIDS.

overgrowth in the upper airways and bacterial toxins were absorbed into the blood stream and this led to sudden death. In the late 1980s it was realized that infants sleeping on their fronts (prone sleeping position) were at increased risk of SIDS. Studies have subsequently shown that when infants sleep prone, secretions pool in the upper airways and bacterial growth and toxin production is enhanced. Campaigns to encourage parents to put their infants to sleep supine (back to sleep) have led to a marked fall in the frequency of SIDS and there are now fewer than 400 cases per annum in England and Wales.

Another risk factor for SIDS is parental smoking, both during pregnancy and after birth. Studies using animal models have shown that nicotine and bacterial toxins can interact synergistically to cause sudden death.

The common bacterial toxin hypothesis is therefore consistent with the age incidence of SIDS, the association with prone sleeping, the seasonal incidence and the risk of exposure to tobacco smoke. Recently staphylococcal toxins have been found in the cerebrospinal fluid (CSF) in cases of SIDS and this provides considerable further support for this idea.

In the 1980s most cases of SIDS were found after an overnight sleep, hence the term "cot death". After the success of the back to sleep campaign in the early 1990s the pattern has changed and more cases are found after a day time sleep or after co-sleeping in the parental bed. One possible explanation is that in prone sleeping infants pooling of secretions in the upper airways and overgrowth of bacteria takes time and the risk of death will increase with length of time in the prone position. In supine sleepers this will not occur. But what has emerged is that in infants who die in the daytime they are often well or have only a minor illness when last seen and subsequently die in a short period of time. The average time from last seen to found dead is of the order of 60 minutes.

Our usual concept of fatal infection is that of a gradual deterioration over time, a process in which we become progressively more unwell until death ensues. Sudden death coming out of the blue is rather more difficult to explain. A possibility, however, is that the toxins that cause SIDS are delivered as part of a transient bacteraemia. Transient bacteraemia is a process in which bacteria enter the blood stream but are quickly cleared by neutrophils which phagocytose and destroy the bacteria. These episodes occur throughout life and in most cases there are no symptoms; but when the bacteria are cleared toxins will be released and in infants who lack protective IgG death could ensue. Episodes of transient bacteraemia last less than 20 minutes. Bacteria can divide in 20 minutes and therefore the total number in the blood stream could double in 20 minutes; if we cannot remove the bacterial load in less than 20 minutes the battle will be lost.

Acute life threatening events (ALTEs) can occur in infancy. These are episodes in which infants suddenly stop breathing, change colour and appear to their carers to be "lifeless". During these episodes there is a fall in oxygen levels in the blood (hypoxia) and the heart rate slows as does breathing. These episodes last less than 20 minutes and in the vast majority of cases the infants spontaneously recover and there are no harmful effects. Infants who suffer episodes of ALTE are at increased risk of SIDS; furthermore the physiological changes in SIDS are similar to those in ALTE at least in the very small number of cases of SIDS who were being monitored at the time of death (about 20 cases worldwide). In these latter cases there is a fall in oxygen level in the blood, a slowing of heart rate and breathing and finally cardio-respiratory arrest and death.

A unifying idea is that episodes of hypoxia occur in infancy. If mild and transient they cause no harmful effect (ALTE), but if severe and prolonged they can lead to death (SIDS). A possible explanation for both is transient bacteraemia with toxin release. Once again theoretical considerations guided by information theory lead to testable predictions. I say again that time will tell if the ideas are useful or not.

Chapter Four

Ageing

Introduction

The four laws of information processing are:

- All information processing systems have a finite capacity.
- There is always a finite chance of error when information is processed.
- Information systems decay at random with time and the error rate rises with time.
- All complex information processing systems need a high level of redundancy to reduce the error rate to a minimum.

These four laws can be used to deduce the way in which the error rate will rise with time. A simple model of this process has been presented in the last chapter.

Consider a single decision making system. It has a finite rate of error $1 - R$, where R is the probability of a correct response.

Let us now assume that the probability of a correct response decays at random with time. The error function is now $1 - Re^{-pt}$, where e is the exponential function, p is a constant, and t is time. The constant p being the rate at which the system decays.

The next stage is to add in redundancy. This can be modelled simply by assuming there are n systems acting in parallel. A mistake will only occur if all systems make the same error at the same time. The error function is now $(1 - Re^{-pt})^n$.

In the information model, making a mistake impairs health and contributes to disease. Thus the error function should describe the way health is impaired and disease develops with age. Furthermore the ultimate mistake leads to death. Figure 3.2 shows a plot of mortality with age for men and women together with the error function. The function fits the observed data.

As argued in the last chapter; the essence of ageing is decay in a complex highly redundant system leading to increased error rate with time. The function simply describes that process and is thereby useful.

Ischaemic Heart Disease

Ischaemic heart disease is one of the commonest causes of death in the western world. The disease is caused by the accumulation of fatty material in the arteries that supply the muscle of the heart. The fatty deposits (atheroma) obstruct the flow of blood and lead to death of heart muscle (a heart attack). This can in turn lead to cessation of heart muscle contraction and death of the patient.

A normal coronary artery has a wall composed of three layers; tunica intima, tunica media and tunica adventitia. The adventitia is composed of connective tissue that is found throughout the body. The media is the thickest layer and is composed of muscle and elastic tissue. The intima is a thin layer of connective tissue and is lined on its inner surface (the luminal surface) by a single layer of cells (endothelial cells).

Atheroma is deposition of fatty material in the intima. Fat is an irritant and leads to local inflammation which heals by scarring (fibrous tissue). Thus a plaque of tissue forms in the intima and narrows the lumen; it is composed of a mixture of fat and fibrous tissue. This is termed an atheromatous plaque. The surface endothelial cells overlying the plaque can be damaged and fall off; this leads to an ulcerated atheromatous plaque. A clot of blood then forms on the ulcerated surface and subsequently adjacent endothelial cells divide and grow to recover the ulcerated area. The atheromatous plaque now consists of fat, fibrous tissue and clotted blood. The last contains fat which causes more inflammation and more fibrosis. In this way the plaque slowly enlarges.

Progressive narrowing of the coronary arteries by atheroma leads to a situation in which there is just enough blood to supply the heart at rest, but not enough blood for exercise. This causes angina; the patient suffers chest pain on exertion. A typical history is onset of chest pain when walking uphill, and cessation of pain when stopping to rest.

If an atheromatous plaque in a narrowed coronary artery undergoes ulceration then the clot that forms on the surface might completely occlude the lumen. This leads to myocardial infarction; the muscle supplied by the artery dies. Myocardial infarction causes severe chest pain at rest. If the patient survives the area of dead muscle is replaced over about six weeks by a fibrous scar.

The process of atheromatous narrowing, coronary thrombosis and myocardial infarction can lead to death because of loss of muscle resulting in heart failure or because of a disturbance of the heart rhythm. In normal rhythm a wave of electrical activity passes over the heart and causes the muscle cells to contract at the same time; hence blood is expelled from the heart and pumped round the body. Lack of blood supply to the heart can disturb the normal electrical wave and the muscle cells then contract and relax out of synchrony; the result is no effective heart beat and the patient will die if the rhythm is not restored. Cardiac massage is designed to keep the heart beating until the rhythm is spontaneously restored. Defibrillators deliver an electrical shock to the heart to bring all the cells to the same state in the hope that synchronous activity will then restart.

Conditions which increase the risk of ischaemic heart disease are:
- Increased blood pressure – hypertension.
- Increased tendency for blood clots to form
- Increased blood fats – hyperlipidaemia
- Diabetes mellitus.
- Smoking
- Genetic factors.

Increased blood pressure damages endothelial cells and this is most marked at points of bifurcation in the arterial tree where turbulence occurs in the blood stream. Damage to endothelial cells will increase the rate at which atheroma accumulates in the intima and increase the complications of atheroma such as coronary thrombosis.

A number of conditions such as chronic infection increase the tendency of blood clots to form and this will also accelerate atheroma and increase the risk of complications. Blood clotting tendency is influenced by both genetic and environmental factors. The latter includes chronic infection as already noted. Blood clotting is influenced by a network of genes and both deleterious and neutral mutations can influence the system leading to both increases and decreases in the speed and degree of clotting.

In general the more fat that is carried in the blood, the greater the risk of atheroma and its complications. Once again both environmental and genetic factors are involved. The composition of the blood is influenced by many genes and both deleterious and neutral mutations will have a role in influencing risk. Paradoxically obesity has less influence on blood fat than often assumed, but the precise composition of the diet does have a role.

Both type 1 and type 2 diabetes mellitus increase the risk of atheroma and its complications. Diabetics tend to have increased blood pressure, increased blood fats and an increased tendency to form blood clots. But these factors do not account for all of the increased risk.

Smoking increases the risk of atheroma and its complications. The most likely mechanism is that smoking damages DNA in all the cells of the body. In people who smoke the rate of mutation during cell division increases and therefore mutations accumulate in somatic cells. If endothelial cells are damaged by the accumulation of deleterious mutations they will be less able to resist the steady accumulation of atheroma, more likely to ulcerate, and slower to restore the lining.

Neutral and deleterious mutations will influence many aspects of atheroma. Neutral mutations influence blood pressure, blood clotting, blood lipids and endothelial cell function. Deleterious mutations interact to impair the performance of networks of genes and will also influence the above four features i.e. blood pressure, blood clotting, blood lipds and endothelial cells.

The simplest way of viewing ageing, ischaemic heart disease and atherogenesis is by consideration of the actions of deleterious mutations. The zygote starts with a number of deleterious mutations and these will interact to impair performance in genetic systems including systems concerned with blood pressure, endothelial cell function etc. But as the zygote divides and its daughter cells divide through 30 or 40 cell generations new deleterious mutations will arise in the progeny cells. There are between a billion and a thousand billion stem cells in our bodies (30 to

Ageing

40 generations from the zygote). All these cells have the set of deleterious mutations in the zygote plus a few more arising during somatic cell division. The total deleterious cell load will impair performance and in the case of endothelial cells lead to increased risk of atheroma. It is important to remember, as stated in chapter one, that deleterious mutations interact synergistically to impair performance. Thus early in life the cells carry deleterious mutations but there is no significant effect on endothelial cell function etc. In children and young adults atheroma is rare. But later in life more mutations occur and impairment in performance becomes significant; atheroma then forms. Individuals who start with the most deleterious mutations will reach critical levels sooner and will appear to age more quickly. But still they will not develop atheroma as children or young adults, only when they reach middle age. Smoking will accelerate the process and those who smoke do appear to age more quickly.

A genetic condition in which the process of ageing appears to accelerate is progeria. This is a recessive disease; that is, individuals inherit two defective copies of the same gene. In progeria children show premature ageing and in their early twenties they resemble elderly people. Men are prematurely bald, have wrinkled skin and develop severe atheromatous disease leading to early death. The gene involved is part of the DNA repair process and these individuals have an increased rate of mutation during cell division. This condition indicates how important mutations are in ageing.

Another recessive disease leads to premature atheroma. In this condition those who have two defective copies of a gene concerned with lipid metabolism have very high blood lipids and develop ischaemic heart disease as children. But they are not prematurely aged in other respects.

The incidence of ischaemic heart disease and its complications has fallen in the western world over the last 20 or 30 years. This means that the risk of say a man of 65 years of age dying of a heart attack is less than it was 20 or 30 years ago. What has happened is that we are living longer and therefore heart attacks still occur but at a progressively later age. Drugs to control blood pressure, blood lipids and blood clotting have certainly played a part. Public health measures directed at stopping people smoking are also important. But in addition other factors, which we do not fully understand, have played a part. A more varied diet might be a factor; certain foods contain anti-oxidants which help to reduce the rate of mutation in cells.

Whatever the reason, we seem to be healthier and to be ageing more slowly.

I will consider this question of temporal changes in ischaemic heart disease in a later chapter as it is relevant to the discussion on health inequality.

Cancer

The common cancers, such as cancer of the lung, colon, prostate, stomach and pancreas rise in frequency with age.

The prevalence of cancer is proportional to age raised to the fifth, sixth or seventh power i.e. t^q where q is equal to 5,6 or 7.

The prevalence of cancer in men at age say 70 is the proportion of men who have developed the cancer by that age. The incidence of cancer in men at 70 is the proportion of men who develop the cancer (or are diagnosed with cancer) at age 70. The incidence is always less than the prevalence. In fact the incidence is the rate of change of prevalence with time and those who remember calculus from school days will know that if prevalence is proportional to t^q then incidence is proportional to t^{q-1} (those who do not remember calculus should just accept it as a fact).

In the 1950s and 1960s a group of mathematicians pointed out that the above relationship between cancer and age could be understood if it were to be the case that cancer was caused by q specific mutations occurring in a single stem cell. The idea being that a set of genes control growth and loss of a specific set would lead a single cell to grow out of control and cause cancer.

In order to try to understand these ideas consider a single cell with genes A, B, C, and D. Let us assume that the chance of a mutation in any one gene is 1 in 100 per year. The incidence of mutation in gene A is thus 1 in 100, but the prevalence will be 1 in 100 for the first year, 2 in 100 for the second year, 3 in 100 for the third year and 4 in 100 for the fourth year, and so on. Thus after four years the incidence is still 1 in 100 but the prevalence is 4 in 100. Exactly the same applies to genes B, C and D. Now let us consider the chance that both A and B have undergone mutation. The incidence of the joint event is 1 in 10,000. The prevalence of the joint event is 1 in 10,000 for year one, 4 in 10,000 for year two, 9 in 10,000 for year three and 16 in 10,000 in year four. To get the prevalence of the joint event we have squared the prevalence of the single event. What is the prevalence of mutations A, B and C? The incidence of the triple event is 1 in a million. The prevalence is 1 in a million in year one, 8 in a million in year two, 27 in a million in year three and 64 in a million in year four. More generally if there are q mutations the prevalence will rise with age raised to the power q.

The idea that cancer is due to q mutations is the simplest model. A more complex model is to postulate that several mutations lead to loss of growth control and the development of a neoplasm (literally new growth), and then further mutations in the expanding pool of cells leads to the formation of a cancer. In this model cancer develops in stages. In this particular model the number of mutations for cancer can be less than q.

Another idea was that each mutation increased the likelihood of the next mutation. Thus the genome becomes progressively unstable, in which case the number of mutations for malignancy could be more than q.

These basic ideas, produced by mathematicians using just pencil and paper, have been shown to be correct. Cancer is caused by the accumulation of specific mutations in a single cell. That cell then grows out of control to produce a neoplasm (new growth). Other cancers arise in stages. For instance in intestinal cancer the first stage is a benign growth (a polyp). The next stage is one of the cells in the polyp acquires further mutations and forms a cancer. Finally in the process of carcinogenesis each mutation contributes to genetic instability and the probability of the next mutation is increased.

The mathematicians who produced these ideas, however, also pointed to a certain paradox. Men are one thousand times larger than mice and live thirty times as long. The probability of cancer developing in a single cell rises as the fourth, fifth or sixth power of age. Thus the probability of cancer per cell should be at least a million times greater in men than in mice (30 raised to the power 4 is 810000; 30 raised to the power 5 is 24300000). Men are 1000 times larger and therefore have a 1000 times more cells; thus the overall risk of cancer should be at least a billion times greater in men than in mice. In fact the prevalence of cancer is similar in mice and men rising to similar rates at the end of their respective lifespan. This is the paradox. Men and mice have a similar number of genes, their cells are similar in size, the mutation rate per gene per cell division is similar, and systems of DNA repair are the same.

A paradox indicates that there is something wrong with our basic assumptions. There is good evidence that cancer arises in a single cell and that it is caused by the accumulation of specific mutations within that cell. So to find the answer to the paradox we need to look more closely at how cells are generated and the way they are related.

Stem Cells

Each individual starts as a single cell, the zygote. The first cell division produces two cells. The next round of mitoses produces four cells. The next round produces eight cells and so on. After 30 rounds of mitoses there will be a billion cells and after 40 rounds a thousand billion cells. The number of cells in the body of an adult male is over ten thousand billion, and approximately a million billion cells are produced in a human lifetime.

The conventional view is that of the order of a billion stem cells are produced from the zygote after 30 rounds of division and then each stem cell maintains an anatomical unit by a process of asymmetric division. Consider a part of the skin made up of 256 cells. The idea is that a single stem cell divides to produce two daughter cells. One daughter becomes the new stem cell and the other goes through 8 rounds of mitoses to produce 256 cells which form part of the skin. A few days later the process is repeated. The new stem cell divides to form two daughters, one becomes the new stem cell and the other goes through eight rounds of division to produce a new set of 256 cells. This set displaces the previous set and forms part of the skin. The previous set is shed from the skin surface.

In this model the original stem cell is approximately 30 mitotic divisions from the zygote. But if the stem cell divides every week thereafter it will be another 3500 divisions from the zygote by age 70 years. With each round of cell division there is a risk of mutation and this neatly explains how the risk of cancer will rise with age.

But it is possible, in theory, to produce all the cells in the body with far fewer than 3500 cell generations. As detailed above, 30 cell generations from the zygote produces a billion cells; 40 rounds of division produces a thousand billion cells; 50 rounds produces a million billion and 60 rounds a billion billion. In fact 60 rounds will produce enough cells for one thousand years of human life. It is therefore possible, in theory, to produce all the cells in the body with no cell, even in extreme old age, more than 60 cell divisions from the zygote.

If there are five specific mutations for malignancy a cell which has undergone 600 divisions will be one hundred thousand times more likely to become a cancer than a cell that has undergone only 60 divisions. Obviously restricting the number of cell generations to 60 will vastly reduce the risk of malignancy.

The model of stem cell generation that restricts the number of divisions to 60 is hereafter referred to as the hierarchical model. It is a more complex system than the conventional stem cell model. The model is illustrated in the table. The primordial stem cell, produced after 30 rounds of division from the zygote is labeled generation A. That stem cell has the task of producing 256 cells per day to populate an anatomical unit. No differentiated cell even in old age must be beyond generation Z. The process is as follows. The stem cell of generation A divides to form two daughters of generation B. One B cell becomes a quiescent stem cell and the other divides to form two cells of generation C. One C cell becomes a quiescent stem cell and the other divides to form two cells of generation D. The process continues until there is a linear hierarchy of quiescent stem cells $BCDEFGHIJKLMNOPQ$ and one active Q which undergoes eight rounds of division to produce 256 cells of generation Y. The generation Y cells are fully differentiated cells that populate the anatomical unit for one day. On day 2 the quiescent Q is activated and undergoes eight rounds of division to produce 256 cells of generation Y which displace the previous set. On day 2 quiescent P divides to form two cells of generation Q. The latter populate the anatomical unit on days 3 and 4. On day 4 quiescent O is activated and forms quiescent P and two cells of generation Q. The latter maintain the anatomical unit on days 5 and 6 and then quiescent P forms

TABLE

A – BB – BCC – BCDD – BCDEE – BCDEFF – BCDEFGG –
BCDEFGHH – BCDEFGHII – BCDEFGHIJJ – BCDEFGHIJKK –
BCDEFGHIJKLL – BCDEFGHIJKLMM – BCDEFGHIJKLMNN –
BCDEFGHIJKLMNOO – BCDEFGHIJKLMNOPP –
BCDEFGHIJKLMNOPQQ

BCDEFGHIJKLMNOPQ..........Q divides through 8 generation
to form 256 Y (day 1)
BCDEFGHIJKLMNOQQ.........Q forms 256 Y on day 2.
BCDEFGHIJKLMNOQ...........Q forms 256 Y on day 3.
BCDEFGHIJKLMNPQQ.........Q forms 256 y on day 4.
BCDEFGHIJKLMNPQ - - - - - - Q forms 256 Y on day 5.
BCDEFGHIJKLMNQQ............Q forms 256 Y on day 6.
BCDEFGHIJKLMNQ..............Q forms 256 Y on day 7.
BCDEFGHIJKLMOQQ............Q forms 256 Y on day 8.

*In this model stem cell A produces a linear hierarchy of quiescent stem
cells labeled BCDEFGHIJKLMNOPQ. Stem cell Q divides on day 1 to
produce 256 cells of generation Y which populate the anatomical unit for
one day. On the second day the second Q divides to produce a further 256
cells of generation Y which displace the first set and populate the
anatomical unit for the second day. On day 2 stem cell P divides to form
two more cells of generation Q. This process continues with a cell of
generation Q producing 256 cells of generation Y daily. P divides for the
first time on day 2 and its progeny maintain the anatomical unit for 2 days.
O divides for the first time on day 4 and its progeny maintain the
anatomical unit for 4 days. N divides for the first time on day 8 and its
progeny maintain the anatomical unit for 8 days. M divides for the first
time on day 16, L on day 32 and so on to B which divides for the first time
on day 32768 (90 years). The anatomical unit will run out of cells after 180
years.*

*Stem cells operating in this way would maintain the anatomical unit for up
to 180 years with no cell more than generation Y (i.e. 55 generation from
the zygote). This model would vastly reduce the risk of cancer compared
with conventional models.*

two more Q cells for days 7 and 8. On day eight quiescent N divides. Quiescent M divides on day 16, quiescent L divides on day 32 and so on. With this system Q maintains the anatomical unit for one day, P maintains the unit for 2 days, O maintains the unit for 4 days, N keeps it going for 8 days, M for 16 days, L for 32 days, K for 64 days, J for 128 days, I for 256 days, H for 512 day, G for 2.8 years, F for 5.6 years, E for 11.2 years, D for 22.4 years, C for 44.8 years, B for 90 years, and A for 180 years. Thus the original A can maintain the anatomical unit for 180 years producing 256 cells per day with no differentiated cell more than generation Y.

If stem cells are organized as suggested by the hierarchical model the risk of cancer is vastly decreased. In fact cancer would be rare in both men and mice as long as the strict progression of the hierarchy is maintained. But the control of the hierarchy would be complex and any errors that occurred would lead to an increased number of cell generations and an increased risk of cancer. The prediction therefore is that cancer will be rare as long as the hierarchy is maintained and will only occur in practice when control of the hierarchy is disturbed or lost. The risk of loss of control in a complex system will rise with age (as in figure 3.1); thus the rising risk of cancer with age is not determined by the number of mutations to cause cancer but by the ageing function of complex systems.

This idea explains the man mouse paradox. Cancer rises to similar levels at the end of the natural lifespan regardless of length of lifespan. It is a product of the ageing function in that species.

In an adult human there are approximately 350 billion cells produced per day. Of these cells 200 billion are red cells, 100 billion are neutrophils and 17 billion are epithelial cells of the small intestine. This means that over half of the stem cells in the body are red cell precursors. If the conventional model of stem cell generation were correct then the majority of mitoses would be in the red cell series and cancer of the stem cells of the red cell series should be common. In fact erythroleukaemia (malignancy of red cell precursors) is extremely rare. This is another paradox for the conventional model. The hierarchical model, however, explains this quite simply. Loss of control of the hierarchy in epithelial cells can lead to a differentiated cell reverting to a stem cell phenotype and taking over the provision of cells for an anatomical unit. This would lead to an increased number of cell generations. This cannot occur in the red cell series because the differentiated red cell loses its nucleus. Thus the number of cell generations in the red cell series is always a low number and malignancy is rare. Furthermore if half the cells in a human can be produced with a very low risk of malignancy it shows that, so long as the hierarchy is maintained, cancer will be rare in man. In the same way it would also be rare in mice.

One of the conventional ideas in medicine is that carcinogens are mutagens. This certainly applies to tobacco smoke and to radiation. But we have long been aware that inflammation can cause cancer. For instance the risk of cancer of the colon and rectum is increased in patients with ulcerative colitis; cancer of the oesophagus is increased in patients with reflux oesophagitis; and cancer of the skin is increased in patients with chronic skin ulcer. Furthermore patients with cancer often report trauma to the area where the cancer has arisen. This is usually dismissed as recall bias; patients without cancer forget the trauma and patients with cancer remember the trauma. In a study in Lancaster we interviewed women with breast cancer and a large number of controls without breast cancer. The women with cancer reported more episodes of trauma to the breast in the five years prior to diagnosis than the controls. The male researchers, including me, judged this to be recall bias. The female researchers said "no way is this recall bias". As always the women were right. Recall bias means that women without cancer had suffered as much trauma to the breast but had forgotten it; my female colleagues assured me that any women who was kicked or punched in the breast would never forget it. The answer of course is that trauma and inflammation will disrupt the hierarchy and it is not only mutagens that cause cancer.

The Quiescent Stem Cell

In the colon there are crypts lined by epithelial cells. Each crypt is the product of a single stem cell. But at the base of the crypt we do not see 15 quiescent stem cells as suggested by the hierarchical model. Thus if the model is correct the quiescent stem cells must reside elsewhere; perhaps in the bone marrow or perhaps amongst the surrounding mesenchymal cells. In either case it makes the concept of hierarchical control even more complicated.

The quiescent stem cell of the immune system is the lymphocyte. It is the memory cell; the cell which has rearranged its genes to produce a specific immunoglobulin or T cell receptor as a result of an encounter early in life, and remembers the encounter into old age. The memory is that the cell is quiescent until it is required, perhaps many years later. The lymphocyte has a condensed nucleus and scanty cytoplasm. Its task is to keep the DNA intact until required. Another interesting property of lymphocytes is that they are very sensitive to radiation damage. The best thing that can happen to a quiescent stem cell that acquires mutations in the resting phase is that it dies. If not it will generate mutant stem cells and mutant epithelial cells with an increased risk of cancer.

Early in the generation of the lymphocyte series T and B lymphocytes rearrange their genes and bring together components at random to produce between 100,000 and 1,000,000 different immunoglobulins and T cell receptors. These individual receptors are then labelled in some way as recognizing self or recognizing not self. The not self are then chosen for clonal expansion when an appropriate pathogen is encountered. The memory of the encounter is an expanded clone of quiescent lymphocytes present for life. The number of cell generations that it takes to generate mature T and B lymphocytes and plasma cells which are involved in inflammatory responses can only be kept below 60 if there is a linear hierarchy as in the case of epithelial cells, the linear hierarchy being composed of quiescent stem cells which resemble mature small lymphocytes.

An Interesting Twist

If cancer is due to the accumulation of q mutations in a single cell then there are situations in which individuals with few deleterious mutations in the zygote could be at increased risk of cancer. This is because in those with many deleterious mutations in the zygote the occurrence of somatic mutation is likely to lead to cell death before an additional q mutations arise. Biology is complicated.

Mitochondria and Ageing

Mitochondria are cellular organelles which generate energy by consuming oxygen. They have their own DNA which is 16,000 bases long. The DNA of mitochondria is similar to that of bacteria in that all the bases are used for coding. It is thought that mitochondria have evolved from bacteria which parasitized cells in the remote past. Mitochondria divide asexually.

Mitochondrial DNA has a higher rate of mutation than nuclear DNA because of close exposure to activated oxygen molecules (remember oxygen damages DNA and anti-oxidants in food protect) and reduced availability of DNA repair enzymes. There are up to a thousand copies of mitochondrial DNA per cell and the mitochondria in muscle cells turn over every 10 days. Thus there are many divisions of mitochondria in a human lifetime. If mutations occur the mitochondria can cease to function but survive as an organelle. In this way cells could accumulate non-functioning mitochondria resulting in cell damage.

The accumulation of mutant mitochondria in long lived brain and muscle cells can lead to degenerative neurological and neuromuscular disease occurring in later life and there is active research into the possibility that mitochondria contribute to the ageing process.

Mitochondrial disease is another example of decay in an information storage system. Asexual reproduction is successful because the information store is small and the level of redundancy is low. Mutations will lead to death of the mutant mitochondrion and it might mean death of the cell in which the mitochondria reside. The mitochondria which are passed from generation to generation through the ova, however, are the ones without significant mutations.

Conclusion

There are a number of other degenerative diseases which rise in incidence with age in the same way as ischaemic heart disease and the common cancers. These conditions include pulmonary emphysema, osteoarthritis and senile dementia. In each case the risk of the disease developing increases rapidly with age just as the risk of dying increases rapidly with age. The simplest way of understanding this process is as decay with time in a complex highly redundant information processing system. The rate of decay, however, is influenced by the environment. Important factors which speed up the rate of ageing include chemical mutagens and radiation which directly damage DNA. But any process which disrupts cellular organization will increase the number of cell divisions and also increase DNA damage.

Those individuals who have a high number of deleterious mutations in their genome will be at a disadvantage from the outset. It will take fewer additional somatic mutations to lead to the diseases of old age and they are likely to get there sooner. They will age more rapidly. The deleterious mutations they carry arose in their parents, grand parents and great parents over many generations. They are a legacy of environmental conditions over the last few hundred years. Environmental conditions have been improving over the last century and there is a good chance that the mutation rate has been falling and that we can look forward to a gradual decrease in the mean number of deleterious mutations in the genome. If so we will all lead longer and healthier lives. But industrial societies pollute and we cannot assume that the mutation rate will continue to fall. There is an urgent need to develop methods for measuring the mutation rate in our cells so that population monitoring can be introduced.

Chapter Five

Health and Disease

Introduction

Most human infants, born into the developed world, can now expect to live into old age. If they survive the first year then death in childhood, adolescence or early adult life is rare. The main causes are infection, cancer or accidents. In later life the degenerative diseases, such as arterial disease leading to heart attacks and strokes, or lung disease leading to progressive respiratory impairment, are major causes of death together with cancer. Infection, autoimmune disease and cancer have been described in previous sections. It has been suggested that all these conditions are best regarded as an interaction between the environment and our genetic constitution:

- Deleterious mutations in the germ line interact synergistically to degrade every aspect of performance and therefore they will impair decisions made in response to environmental challenges.
- Neutral mutations in the germ line will impair the response to specific environmental challenges but will enhance the response to others; overall the effect is neutral.
- Deleterious mutations also arise during somatic mutation; they gradually accumulate in our stem cells and are a key part of ageing. As our stem cells lose their information store they become less able to respond to environmental challenge.
- Environmental challenges include micro-organisms, chemical pollutants, radiation, physical factors such as trauma, and harmful effects of too much or too little food.
- The situation is even more complicated when we remember that environmental mutagens increase somatic mutation and lead to an increased number of deleterious mutations in the germ line of future generations.

The key public health messages are that one cannot improve health by manipulating the genome, and we should not try; but we can improve health by modifying the environment and that extends to reducing deleterious mutations in future generations.

Members of the medical profession, particularly pathologists such as me, are apt to define disease as the absence of health and to define health as the absence of disease. But in recent years there has been a move, with which I reluctantly concur, to regard health as more than the absence of disease. One can be unhappy, overweight, physically unfit, live a socially and emotionally impoverished life and yet be free of a medically diagnosable disease. Free of disease but not in the best of health. I would contend, however, that we can analyse health as we have analysed disease. Deleterious mutations will impair performance of a wide range of capabilities and that will extend to making friends, developing social relationships, maintaining an appropriate body weight, developing intellectual interests, artistic appreciation etc. Neutral mutations will influence aspects of personality; after all the regulation of cytokine response to infection involves the same molecules that influence the flow of information in the brain and hormonal responses in the body. Finally the interaction between the environment and our genes will determine the life we live and the degree of health we enjoy.

In this chapter I intend to examine some aspects of health and disease not covered in previous chapters. The aim is to show how the ideas developed in previous chapters can be used to generate hypotheses and to provide explanations for a range of different conditions.

Schizophrenia

Schizophrenia is a mental disease that occurs in 1% of the population at some time in their life. It first presents in adolescence or early adult life and is characterised by thought disorder. The condition is equally common in men and women. The cause is unknown but there is evidence of a genetic component. If one of a pair of identical twins develops schizophrenia there is a 60% chance that the other twin will develop the disease. For non identical twins the risk is closer to 10%.

Schizophrenia is unusual in that the prevalence of the disease is similar in a wide range of different countries and different racial groups. The lifetime prevalence is close to 1% regardless of race, geography, culture or state of technological advance.

If a single mutant gene causes a disease then 50% of the children of an affected parent will carry the gene and develop the disease; this is dominant genetic disease but it is not the pattern seen in schizophrenia. If deletion of both copies of a gene leads to disease then 1 in 4 siblings of an affected person will develop the disease; this is recessive disease but again it is not the pattern seen in schizophrenia. The genetic mechanism in schizophrenia is thought to be polygenic; this means several mutant genes together cause the disease.

The human genome has 25,000 loci with two alleles (genes) per locus; one allele from mother and one from father. The alleles at a particular locus can be the same or different. Differences, if present, are due to changes in base sequence and are known as polymorphisms. The majority occur in control sequences rather than in the main protein coding sequence. If a particular locus has two genes A and B the possible genotypes are AA, AB, or BB. In practice genes A and B will be found in all racial groups but the frequency will vary and much of the variation is purely a matter of chance. Thus if a disease is due to several genes at several different loci one would expect the prevalence of the disease to vary markedly between different racial groups. A constant rate as in schizophrenia is unusual.

One possible explanation is that the genes that lead to schizophrenia also confer some advantage. Consider three genetic loci with genes A or a, B or b, and C or c. Let us assume that if an individual has genes a, b and c then they are at risk of schizophrenia (this occurs in genotype Aa, Bb, Cc). Individuals who develop schizophrenia are less likely than the general population to have children so there is a selective disadvantage against genes a, b and c. But if possession of one or two of the genes a, b, or c confers a procreative advantage then individuals with genotype AA, BB, CC will be at a disadvantage. Heterozygotes with genotype Aa, Bb, CC or AA, Bb, Cc or Aa, BB, Cc will produce more children than those with AA, BB, CC and the genes a, b, and c will be maintained in the population. This could apply to all populations regardless of geography, culture, race or technological stage.

There are a number of counter arguments to the above explanation:

- The explanation predicts that the non schizophrenic relatives of patients with schizophrenia will have more children than the general population. There is no evidence to support this.

- A considerable amount of research has been undertaken looking for the genes for schizophrenia. The disease is common (1% lifetime prevalence) so there are many families to investigate. If there were as few as three genes that cause the mental disorder they would have been found by now. If there are many more than three then the risk in relatives would be less than 10% (this is a complicated argument but the more genes involved the closer the risk in relatives is to the general population risk of 1%).

- The most important argument, however, is that the idea that there are genes for a disease is a misconception. Schizophrenia is a disease in which there is thought disorder. It is a mistake to imagine that there are genes which cause thought disorder; if they existed there would be strong selective pressure against them and they would disappear. There are genes that specify the systems which allow thought to occur, and mutations in those genes will impair the process of thinking. Thus genes for thought disorder are in fact mutations of genes for thought.

- Thought is not specified by just one or two genes. It is the product of very many genes; indeed it is possible that most of our genes are involved directly or indirectly in creating the systems for thought. Deletion of any of these genes will impair thought and increase the risk of thought disorder. Thus in seeking an explanation for the genetic basis of schizophrenia it is necessary to reconcile the need for a few mutations, in order to explain the high recurrence risk in siblings, with the competing idea that very many genes must be involved. As before, the concept of the "synergistic interaction of heterozygote deletions" provides an answer to the dilemma.

The hypothesis is that there are hundreds or thousands of genes specifying the aspect of thought that is disturbed in schizophrenia. If all the genes function normally the risk of schizophrenia is extremely low. If a single gene is deleted by mutation then the risk of disease remains low because the system is highly redundant. The loss of any two genes raises the risk, and the loss of any three genes raises it to 60%. The loss of four genes raises the risk of malfunction to such a level that the zygote carrying those mutations will abort.

Let us assume that there are several thousand genes specifying a system concerned with an aspect of thought. The mean number of deleterious mutations in this system is 0.6 per person. Three mutations will occur in this system in 1.9% of the population (this is calculated using the Poisson distribution). If 60% of these individuals develop schizophrenia the lifetime risk in the general population will be 1.14%. Approximately 19% of siblings will also have three mutations and 11.4% will develop the disease.

This model gives a simple and plausible explanation for a number of aspects of schizophrenia;

- There is no evidence that the rate at which new deleterious mutations arise differs markedly in different populations and therefore the model predicts that the prevalence of the disease will be similar in different communities.
- There are several thousand different deleterious mutations that can contribute to schizophrenia even though in any one person only three are involved. This is why the schizophrenia gene has not been found and never will be found.
- An intrinsic part of the model is that complex systems make decisions in an uncertain world. The presence of deleterious mutations increases the risk of error but does not guarantee error. The genetic mutations do not mean that disease is inevitable and the environment has an important role.
- Schizophrenia concerns an error or failure of a single complex system and in that sense is homogenous. The genetic basis, however, is markedly heterogenous and it is not surprising that the phenotype is variable.
- Alternative theories predict a procreative advantage for the relatives of schizophrenics for which there is no evidence. The model based on deleterious mutations does not need this assumption.

There are many human diseases that are thought to be polygenic in that there is evidence of a genetic cause but the data do not fit with dominant, recessive or sex linked patterns of inheritance. Putative polygenic conditions include ischaemic heart disease, hypertension, stroke and maturity onset diabetes mellitus. In fact most disease is a consequence of an interaction between the environment and our genetic constitution; in a minority of cases there is a simple genetic pattern but in most the genetic component is polygenic. Whenever polygenic disease is under investigation, the possibility of a causal mechanism based on the "synergistic interaction of heterozygote deletions" should be considered.

The Barker Hypothesis

Professor Barker and his colleagues from Southampton have shown that low birth weight babies are more likely, in later life, to develop hypertension, ischaemic heart disease and maturity onset diabetes mellitus than are larger babies. His group noticed that the geographical distribution of risk for ischaemic heart disease in England and Wales in the second half of the twentieth century was similar to the distribution of increased perinatal mortality in the first half of the century. This led them to examine birth records of infants and relate these to disease incidence in later life.

In Sheffield they found that the standarised mortality rate (SMR) for ischaemic heart disease was 119 for men who had weighed less than 5.5 pounds (2500 gm) at birth. Mortality rates are standardised so that the average is 100 for the country as a whole. Male infants in Sheffield who weighed over 8.5 pounds (3860 gm) had a SMR for ischaemic heart disease of 74. Thus there is a 50% increase in risk across the birthweight range.

The changes are even more marked for maturity onset diabetes mellitus (type 2 diabetes mellitus). In one study of men aged over 64 years with type 2 diabetes mellitus, hypertension and raised blood lipids the prevalence varied eighteen fold with the birth weight range. Men who weighed less than 5.5 pounds at birth were eighteen times more likely to have the above syndrome than men who were over 9.5 pounds at birth.

Similar results are found in women but the studies are more difficult to conduct. The research process involves examining a large number of birth records and then tracking down the individuals in later life to see what has happened to them. Since men retain their surnames they are much easier to find than are women. Furthermore unmarried women are easier to locate than are married women and this introduces a possible bias.

Professor Barker's group has found the above results consistently in a large number of studies and the link between birth weight and increased risk of disease in later life is well established. The explanation for the link, however, is not agreed. There are three possibilities:

- Birth weight is determined by genetic factors and the same genetic factors predispose to disease in later life.
- The environment *in utero* programmes the infant's metabolism and physiology, influencing birth weight and influencing disease in later life. This is Barker's hypothesis.
- The social circumstances that lead to low birth weight also act throughout life and predispose to disease in later life. For instance if parents smoke infants are more likely to be small at birth; the same infants are more likely to take up the smoking habit in later life and be at increased risk of disease.

Thus the cause of the link could be genetics, the environment *in utero* or the environment after birth. As a general rule in biology if there are three broad possible explanations it is more likely that all three apply to a varying degree than that just one applies.

The Barker hypothesis is best stated in Barker's own words "Poor nutrition, health and development among girls and young women is the origin of high death rates from cardiovascular disease in the next generation. It prejudices the ability of mothers to nourish their babies *in utero* and during infancy. The fetus responds to undernutrition with permanent changes in its physiology and metabolism, and these lead to coronary heart disease and stroke in adult life".

Changes over Time

Type 2 diabetes mellitus is due to a relative lack of insulin. It is more common in those who are obese. In primitive communities where food is scarce type 2 diabetes mellitus is rare. In newly affluent communities when the previously poor gain access to plentiful supplies of food type 2 diabetes mellitus becomes common, as does obesity. In the established affluent, those who have had access to plentiful food for several generations, type 2 diabetes mellitus has an intermediate prevalence. Thus over several generations the disease can go from being rare (<1% life time prevalence) then can rise rapidly to being very common (> 30% prevalence as in Polynesia) and then fall gradually to being just moderately common (5% prevalence as in the affluent West). Ischaemic heart disease has shown a similar pattern, rising in incidence in the newly affluent and then falling in the established affluent.

One explanation put forward to explain changes over time was the "thrifty genotype hypothesis". The idea was that for most of our history the human condition has been one in which food was scarce and there has been selection for genes that encourage avid storage. These genes are advantageous in times of food scarcity but become disadvantageous in times of excess. Thrifty genes lead to obesity when food is plentiful predisposing to type 2 diabetes mellitus and ischaemic heart disease. This idea explains the rise in cardiovascular disease with the onset of affluence but it does not explain the subsequent fall in the established affluent. There could be selection against genes for avid storage in affluent communities but it would take many generations before the prevalence changed, particularly since cardiovascular disease arises after the reproductive years.

The Barker hypothesis, by comparison, can explain both the rise with the onset of affluence and the subsequent fall. The idea is that when food is scarce infants are programmed *in utero* for scarcity. If they subsequently grow up in a world of scarcity their metabolism is appropriate for their circumstances and disease is avoided. But if they enter a world of plenty their metabolism is inappropriate and avid storage leads to obesity and disease. For subsequent generations, however, programming *in utero* is designed to anticipate a world in which food is plentiful; storage is less avid, obesity less likely and disease less common.

Problems with the Barker Hypothesis

A distinction should be made between the Barker phenomenon and the Barker hypothesis. The Barker phenomenon is that there is a link between birth weight and disease in later life; there is a substantial body of experimental support for this and it is established beyond reasonable doubt.

Health and Disease

The Barker hypothesis is one possible explanation of the link; it could be correct in part, in whole or not at all.

There are mechanisms by which fetal metabolism and physiology could be programmed *in utero*. Chemical groups can be added to gene regulatory elements thereby switching off or switching on whole banks of genes and changing the way in which the infant responds. The idea that the ideal environment could programme the optimum response pattern and avoid disease is attractive; and there is currently a great deal of research in which the aim is to identify what constitutes that ideal.

In Holland, at the end of the Second World War, there was a severe famine. Barker and colleagues have studied the effects of the famine on infants *in utero* at that time. They discovered, as predicted by the hypothesis, that the infants were at increased risk of type 2 diabetes mellitus, but the effect was small. Equally the famine did lead to decreased birth weight, but once again the effect was very small. The mean birth weight for infants born prior to the famine was 3388 gm, for those conceived after the famine it was 3450 gm; for those *in utero* during the famine it was 3166 gm (late gestation exposure to famine), 3220 gm (mid gestation exposure), and 3436 gm (early gestation exposure). If severe famine has such a small effect on birth weight and risk of disease it is difficult to see how more mild environmental changes could have a larger effect.

In a study conducted in Lancaster and Kendal of 2350 singleton live births environmental factors had only a small effect on birth weight. Maternal smoking was associated with a 266 gm fall in mean birth weight, the mean birth weight change across the social classes was only 165 gm, the change in mean across the seasons was only 42 gm. The Barker phenomenon is that the standardised mortality rate rises by an order of 50% over a birth weight range of approximately 1500 gm. Big changes in the environment lead to only small changes in birth weight and it seems unlikely that the link between birth weight and disease can be explained by environmental effects *in utero* alone.

Standard textbooks of genetics give lists of concordance rates for disease in monozygotic (identical) and dizygotic (non identical) twins. Identical twins develop from the same zygote and have identical genes. Non identical twins develop from different zygotes and are as close genetically as non twin siblings; they share 50% of their genes. In schizophrenia the concordance rate in monozygotic twins is 60%, the concordance rate in dizygotic twins is closer to 10%. This indicates a strong genetic component in schizophrenia as described in the previous section. The fact that the concordance rate is not 100% indicates that the environment is also important. The concordance rate for type 2 diabetes mellitus in monozygotic twins is 100% and dizygotic twins 10%. This would usually be taken as extremely strong evidence for an underlying genetic mechanism. The proponents of the Barker hypothesis argue that the true concordance rate for type 2 diabetes mellitus in identical twins is less than 100%. They do have a point, in that establishing a 100% concordance rate is difficult, as type 2 diabetes mellitus can arise late in life and individuals can die before they develop the disease. But even then the concordance rate is high and points to a strong genetic component.

The Synergistic Interaction of Heterozygous Deletions

The Barker phenomenon can be explained on the basis of the synergistic interaction of heterozygous deletions as set out in previous sections:

- The number of heterozygous deletions in the genome of individuals varies (an approximate Poisson distribution). Those with a few deletions are likely to develop normally and fulfil their genetic potential. Those with more deletions are more likely to be impaired in development. This impairment will take many forms but one is impaired development *in utero* with a smaller and more variable body weight. If infant body weight correlates inversely with the number of deleterious mutations then body weight will also correlate inversely with impaired health in later life.
- Diabetes mellitus and ischaemic heart disease are rare in primitive communities when food is scarce. The onset of affluence leads to obesity and the development of these diseases in later life. The onset of affluence is also associated with improved living conditions, a better all year round food supply and more varied diet with enhanced anti-oxidants and other protective agents (minerals, vitamins etc.). The result is a fall in the new mutation

rate (P) and an eventual fall over several generations in the human genomic load (Q). A fall in Q leads to a fall in the prevalence of diabetes mellitus and ischaemic heart disease.

- This explanation fits with the high concordance rate for type 2 diabetes mellitus in monozygotic twins. Type 2 diabetes mellitus is a polygenic disease, like schizophrenia. A large number of genes are concerned with systems that control blood sugar and lipids. The system is important therefore highly redundant. Individuals with several mutations in the system are liable to develop the disease. Monozygotic twins have the same number of mutations in the system and are at the same risk. Dizygotic twins have a 20% chance (approximate) of having the same number of mutations in the system and therefore a lower concordance rate.

- An increased number of heterozygous deletions is associated with an increased risk of degenerative diseases such as hypertension, ischaemic heart disease, type 2 diabetes mellitus and obstructive lung disease; but a decreased risk of certain types of cancer as discussed above. This predicts that low birth weight will decrease the risk of certain cancers; a somewhat counter-intuitive prediction but this has in fact been found by Barker's group. The Barker hypothesis did not predict this result but *post hoc* modifications can always be made to accommodate such findings; the suggestion being that healthy well fed infants have more cells and therefore are at increased risk of cancer.

Barker's hypothesis is one of the big ideas in medicine. It has led to a vast amount of research and the protagonists of this view have high hopes of influencing the *in utero* experience and improving health. At the same time, running in parallel, but published in different journals, addressed to a different audience and discussed at different conferences, the big ideas in evolutionary biology relate to deleterious mutations and their role in sexual and asexual reproduction. Deleterious mutations arise in mitosis, enter the genome, reduce fitness and cause disease. Models indicate that some form of synergistic interaction must occur to prevent progressive build up and ultimate species extinction.

Specialisation in science means these big ideas, mutually contradictory, can coexist.

Health Inequality

There has been a marked improvement in the health of those living in what is regarded as the developed world over the last 250 years. Industrialisation has brought increased wealth and this in turn has led to an improved supply of food, better transportation, hygienic living conditions, protection from the environment, immunisation and antibiotics. The result is that we now live longer and healthier lives and are much less likely to succumb to the childhood infections that commonly led to death in previous era. The difference of course is due to modification of the environment because our genetic constitution changes much more slowly.

There is a wide gap between the health of the developed and the underdeveloped regions of the world. This is simply because the environment of the developed world has changed but that of the underdeveloped has not. In Africa there is still famine, lack of public sanitation, lack of immunisation, shortage of medical facilities etc. Childhood rates of mortality are high and life expectancy is less than in the developed world. Once again this is health inequality determined by the environment.

When communism collapsed in Russia the economy suffered and there was a sudden and dramatic decrease in the wealth of the majority of the population. This had profound consequences for the health of the Russian population and life-expectancy fell. Once again this is a clear example of the role of the environment in influencing health. The underlying genetic constitution of the population had obviously not changed, although the rate of somatic mutation might well have increased.

What is less obvious, and less easy to understand, is that health inequality is also found within the developed world. Those at the top of the socio-economic scale enjoy better health and live longer than those at the bottom. This is seen most clearly in the British Civil Service. The position of each employee in the hierarchy is defined and this determines their economic and social position in society. Those at the top of the hierarchy are paid more, are less likely to take time off work because of illness, are more likely to survive to pensionable age, and receive their pensions for longer because they survive for longer. Part of the explanation is smoking. Those at the top are less likely to smoke and those at the bottom are more likely to smoke. But even when this is allowed for there is still a gradient of health advantage from top to bottom amongst those who do not smoke.

Food is plentiful in our society, perhaps too plentiful, but the quality of diet can still vary. Those lower down the social scale are more likely to smoke as already noted. Housing conditions vary, with damp and cold not conducive to health. But none of these explanations seem enough to explain the health inequality observed in the developed world and it is tempting to assume that the environmental influences are rather more subtle than those that distinguish the developed from the underdeveloped world.

One possibility is that position in the hierarchy influences our concepts of self worth and this in turn influences health. The suggestion that psychological factors can influence physical disease is neither new nor surprising and it certainly is not unscientific. The molecules that mediate the flow of information in the brain are the same as those that control the action of cells in the rest of the body including those concerned with protection against disease. The relatively small number of genes in our genome (25,000) specify the large number of proteins (over 250,000) in our proteome. The genes and the proteins are found in the brain and in the rest of our body. In fact similar genes and proteins are found in the rest of creation including our commensal flora. We should not be surprised that the brain influences the body nor that the body influences the brain.

Decision Theory

The concept that our sense of self worth can influence health and disease is in fact a direct prediction of information theory or more specifically decision theory. An ideal detector or decision maker is one that uses an optimum strategy to make decisions in an uncertain world. An optimum strategy takes into account *a priori* probability, information and the values and costs of right and wrong decisions. Let us consider a rather contrived example:

An individual named George is placed in a large room with a screen at one end behind which are two doors leading to two separate rooms. In one room is a man-eating lion; in the other room is a man with a cheque for a million pounds made out to George. This is a game but it is played for real. George is told that a card will be chosen at random from a pack. If the card chosen is a spade then the lion will be released. If any other card is chosen then the man with the cheque will be released. George will not know which card has been chosen and which door is opened but he will have a few minutes to listen as the animal (man or lion) walks slowly behind the screen. If he waits until the animal emerges from behind the screen it will be too late to escape and if it were the lion that was released he will die. If he decides to leave then he forfeits the chance of receiving a million pounds.

In this game the *a priori* probability is 0.25 that the lion is released and 0.75 that the man is released. The evidence is determined by listening carefully for the sounds behind the screen. The cost of remaining if the lion is released is death. The value of remaining if the man is released is the prize money. In this game the optimum decision depends on the value that George places on his life, his concept of self worth. If George is a happily married man with a good job and a good income he will leave quickly. A million pounds will only make a small difference to his life, improving it marginally but not much more. His life to him is worth more. But if George is a homeless tramp living on the streets in the depth of winter he might decide it is worth the risk. A million pounds could transform his life and make it worth living.

Biological organisms make a vast number of decisions every day. Those decisions will follow an optimum strategy because evolution will have selected out those that follow any other strategy. Consider the immune system analysing a short polypeptide chain sitting on the surface of a macrophage. Is it self or is it not self? Failure to realise it is not-self could mean a failure to attack a pathogen and could lead to infection and death. Failure to realise it is self could lead to autoimmune disease. The costs of wrong decisions include disability and death; thus self worth is part of the equation. If self worth does not influence the decision then the strategy is not optimal.

How can a concept of self worth influence a T lymphocyte? The answer is we do not know but it must. However although we do not know, there are obvious ways in which the brain can influence the secretion of cytokines by lymphocytes and macrophages and thereby modulate the decision process.

Genetics and Health Inequality

There is strong evidence that the environment in its broadest sense influences health and disease and therefore a large part of health inequality both between and within individual nations is due to environmental differences.

The possibility that genetic differences within a nation could explain part of health inequality is often considered but usually dismissed. The problem with a genetic explanation is it appears to provide an excuse for inaction. If those with good genes rise to the top of the hierarchy and are healthy while those with bad genes fall to the bottom and suffer poor health then there would be little that could be done to reduce the health gradient. It is surely better, so the argument goes, to assume that the environment matters and seek to modify it where appropriate.

The political and social approach to health inequality is often concerned with the redistribution of resources. Take from the rich and give to the poor in the hope that it will benefit the health of the poor more than it damages the health of the rich. At one level this works because the marginal benefit of one pound to the poor is greater than the marginal loss of one pound from the rich. But at another level it can fail because redistribution of wealth can be a disincentive to competition and the overall economy might suffer. If this occurs everybody will lose. The health gradient might be less steep but everybody is worse off.

In this book it has been argued that deleterious mutations in the germ line follow a slightly skewed Poisson distribution. Furthermore deleterious mutations interact synergistically to impair all aspects of performance. Those with fewer deleterious mutations will be more intelligent and better at fighting disease and preserving health. They will rise to the top of the hierarchy and enjoy the economic and social fruits of society. Those who are unlucky and inherit more deleterious mutations will lose out socially, economically and in terms of health. This idea predicts that health inequality in a developed nation will be more about genetic differences than the simple distribution of economic resources.

If an important part of health inequality is due to a Poisson distribution of deleterious mutations then the answer to health improvement is to seek to reduce the rate of mutation during cell division. A better environment with fewer mutagens will mean fewer mutations in succeeding generations and better health for the entire population. Moreover because the deleterious mutations interact synergistically the harmful effect is most marked at the upper end of the distribution. Thus a reduction in the average number of deleterious mutations in the genome will lead not only to better health overall but also to a less steep health gradient in society. Health inequality will not be abolished but it will be less marked.

Those who believe, for the best of motives, that health inequality is due to environment not genetics are led to advocate increasing wealth to improve health; but this means more competition in society and more inequality. But if health inequality is due to deleterious mutations the answer is to reduce the mutation rate and at the same time improve health and reduce inequality.

The Japanese have less health inequality and better health overall than many countries in the West. The average IQ in Japan is also higher than in the West. The prediction based on the ideas presented in this book is that the mutation rate in Japan is less and the mean number of deleterious mutations in the genome is less than in the West. Time will tell; but this is a specific prediction on which to close this chapter.

Chapter Six

Sexual Attraction

Introduction

Many animal species have elaborate and stereotyped mating rituals. Males perform and females choose. But what exactly are females choosing? What are the criteria that determine the acceptable male? Are females choosing good genes or avoiding bad genes? Or is the choice about economics rather than genetics? Or is it love and what is that? Or is it lust and what is that?

Whenever we seek to explain complex behavioural patterns the first question to ask is "is the behaviour determined by genetics or by the environment?" The answer is always the same "genetics and the environment interact to determine behaviour". It is never, or hardly ever, one or the other; it is both. In the case of sexual attraction there will be genetic factors and environmentally determined factors.

The next question is, "if genetic factors are involved in sexual attraction, are we concerned with deleterious mutations or neutral mutations?" In the case of sexual attraction, I think that deleterious mutations are probably the more important, but once again both are involved.

Deleterious Mutations

A key theme in this book is the idea that humans are complex organisms specified by genes but modified by experience. We all carry some deleterious mutations and they interact synergistically to degrade performance. The mean number of deleterious mutations in the germ line of adults is Q but the actual number in any one individual varies (a Poisson distribution) so that some have fewer than Q and some have more than Q. Individuals with fewer than Q germ line mutations will be more intelligent than average, they will be healthier than average and they will be more likely to succeed economically. Furthermore their progeny will have fewer mutations than average; they will be more likely to survive as zygotes, to become adults, and to continue the cycle of success.

There is a clear biological advantage to individuals, male or female, who can recognize and who are attracted to members of the opposite sex with fewer than average deleterious mutations. Their progeny will be more successful. Could this be the basis of sexual attraction?

Let us first of all consider the opposite i.e. that sexual attraction is unrelated to the number of deleterious mutations in the genome. It is possible to develop computer models in which males and females with varying numbers of deleterious mutations mate and produce offspring. The mean number of new deleterious mutations per generation is P, the mean number of deleterious mutations in zygotes is P + Q, the mean number of deleterious mutations in the germ line of adults is Q and the proportion of zygotes who survive to become adults is Zs. In humans the value of P is approximately 1, the value of Q is approximately 6 to 8, and Zs is in the region of 50% to 75%. If P rises then Q will rise and Zs will fall. If P rises too far then Zs will fall to a level at which the population would become extinct.

Let us now consider a polygamous mating population in which females are attracted to males with a few deleterious mutations. If the females are particularly choosy and will only accept the top 10% of the male population (i.e. the 10% of males with the fewest deleterious mutations) as partners then the mean number of deleterious mutations in their progeny will fall. In this situation a high value of P is still compatible with a moderate value of Q and the population will survive.

A monogamous system in which males and females are mutually attracted by low deleterious mutations is intermediate between the polygamous system and the system based on mating without regard to deleterious mutations. For a given value of P random mating (that is mating without regard to deleterious mutations) leads to the highest Q and lowest Zs, polygamous mating leads to the lowest Q and the highest Zs and monogamous mating is in between. Thus a species in which sexual attraction is related to the level of deleterious mutations in the zygote is more likely to survive than one in which there is no such relationship. A polygamous system is required when P is high but a monogamous system suffices when P is low.

For the female there is a balance to strike between the genetic advantage of the polygamous system and the economic advantage of the monogamous system. It might be better to accept a male with a few more deleterious mutations if he will be around to help raise the young. The balance, however, will depend on P. If P is high the polygamous system is best; if P is low then the monogamous system is better.

How Do We Recognize Low Levels of Deleterious Mutations in Members of the Opposite Sex?

In previous chapters it has been argued that individuals with low levels of deleterious mutations will be more intelligent, will be healthier, and will be more successful. These are all qualities and attributes that are easily recognizable and are generally found to be attractive. Clever, witty, successful, able individuals; whether male or female, are invariably attractive to members of the opposite sex. But what about physical beauty? We are undoubtedly attracted by the beautiful body and the beautiful face. What have they got to do with deleterious mutations?

In the 19th century Galton undertook a series of experiments on facial images. His aim was to produce the image of the archetypal thief or murderer by merging the pictures of a number of thieves and murderers. But no matter how ugly, or unattractive or unsavoury the initial images the final merged image was always very attractive and the opposite of what Galton had expected. He discovered that the average face was attractive.

The process of fusing images produces a symmetrical face and it is symmetry that we find attractive. Most faces deviate from perfect symmetry and the more they deviate the less attractive we find them.

During development our genes are used to specify the processes that lead to symmetrical body structures. If there are deleterious mutations then the process will be impaired to some degree and deviation from perfect symmetry will occur.

Thus deviation from symmetry will correlate with the number of deleterious mutations in the genome, and physical beauty will be affected as a consequence.

Disease also impairs physical symmetry and the tendency for disease to occur is related to deleterious mutations as argued previously. Thus there are a number of ways in which physical beauty can be a marker for the number of deleterious mutations in the genome.

A Beautiful Face or a Beautiful Brain?

In order to survive a complex organism has to be capable of performing a range of functions and skills at a high level. For instance a squirrel needs to be able to climb and glide through the trees. Performance on these tasks will correlate with the number of deleterious mutations in the genome and therefore the female squirrel should choose a mate who demonstrates a high level of competence at the task. By comparison physical beauty might correlate with the number of deleterious mutations in the genome but it has no intrinsic value in survival terms. So why is physical beauty such a strong force in sexual attraction? We ought to be attracted by the beautiful brain rather than by the beautiful face, but are we? Answering as a superficial male I would have to say "no comment, as the truth could be incriminating".

"Beauty is only skin deep", is another way of saying the same thing. We should choose on function rather than aesthetics because function matters in the long term.

This argument, however, has an interesting twist. An essential property of a complex highly redundant system is that defects in one or a few components will have no measurable effect on function. Redundancy is the property of having components in excess so that function is preserved if a few components are lost. Thus if sexual attraction is based on function it will not be possible to distinguish between no deleterious mutations and one or perhaps two deleterious mutations.

Now let us think about beauty in terms of information theory. Step one is to take a digital photograph of a beautiful scene in nature so as to produce a beautiful image. The image consists of a large number of pixels each of varying colours. The image has high information content as we would have to specify the colour in every pixel to reproduce the image precisely.

Random changes would decrease the information content and make the image less attractive.

In the case of the peacock's tail, a complex system of genes is required to specify the beautiful structure. If there are deleterious mutations in the system the chance of errors occurring in development will increase and the tail will be less attractive. Since there is no intrinsic functional value to the ornament, in fact it is an encumbrance; there is an advantage to the species if there is a perceptible difference between no deleterious mutations and one or two deleterious mutations.

Consider a lighting system composed of a large number of bulbs which illuminate a stage. The system is highly redundant in that the bulbs overlap and each part of the stage is lit by at least three different bulbs. The key functional requirement is to keep the stage lit at all times. If a few bulbs burn out it will have no effect, but if many fail part of the stage will be in darkness. The difficulty is one cannot tell how many bulbs are working by looking at the stage i.e. assessing function. Now let us add a series of coloured wheels above the lights which project a symmetrical highly attractive moving pattern of coloured lights onto the ceiling. If one bulb goes out the pattern will be slightly less attractive. If several bulbs blow the pattern will be much less attractive. The best way to assess the system is not to look at the stage but to look at the ceiling, not to assess on function but to assess on aesthetics.

Beauty is more than Skin Deep

Function has intrinsic value in terms of survival. Aesthetics has no intrinsic survival value but is probably more important in sexual attraction, the reason being that the best way to signal deleterious mutations is not by impairing function but by impairing beauty.

This will only work in practice if the genes that code for beautiful structures are the same genes that code for function. Beautiful structures signal low levels of mutations throughout the genome because all genes are concerned with function and with beauty.

Mice and men have similar genes. The vast majority of functional genes in men are functional in mice with just minor base changes. The difference between us is related to regulation. Men and mice use the same genes in different combinations to do different things. The genes that specify a mouse both functionally and aesthetically are the same genes, but in different combinations, that specify men functionally and aesthetically. It is not therefore difficult to accept that the genes that specify beautiful structures in men are the same genes, but in different combinations, that specify function.

The conjecture therefore is that every gene in our genome is part of one or more systems with an aesthetic purpose and part of other systems with a functional purpose.

If this analysis is correct then all sexual animals will have some aesthetic appreciation. They will recognize beautiful things and be attracted to beautiful things. Furthermore their appreciation of beauty will be similar to ours; it will be based on complex, information rich, error free patterns. There is a biological imperative to recognize and be attracted to such features, otherwise species will accumulate deleterious mutations and become extinct. The appreciation of beauty could be one of the deepest concepts in nature, something we have in common with all sexual species.

Neutral Mutations

In many species, including our own in more primitive times, there is a polygamous pattern of mating. Just a few males sire the progeny for the next generation. If sexual attraction was based on just one or a small number of neutral mutations then those mutations would be at a marked selective advantage. They would quickly spread through the population and displace the original genes. As a result the basis of the selection would disappear; after a few generations all male would be homozygous for the neutral mutations (homozygous means they would have two copies of the neutral mutation). This is known as the paradox of the lek. In some species of birds the mating ritual involves males gathering in one place and performing their various rituals. The females then gather and make their choice of mate. If the choice was based on just a few neutral mutations present in a sub-set of the males within a few generations all the males would be the same and the basis for choice would disappear.

Let us consider a hypothetical case: the brightness of plumage is determined by a single gene B, and all the birds in the population are homozygous BB. A mutation occurs in one gene in one bird to give a variant of B which we label B. The male BB has brighter plumage than the male BB and sires more progeny. As a result the frequency of gene B increases in the population and it is found in both males and females. Eventually some homozygous males BB will arise, if the homozygote BB is more attractive than the heterozygote BB then gene B will continue to increase in frequency and eventually B will displace B from the population and all birds will be BB homozygotes. In this particular example B is an advantageous mutation rather than a neutral mutation as it has quickly displaced B.

We can be confident that sexual attraction cannot be explained in this simple way. If a few genes for gaudy feathers or large sexual ornaments determined sexual attraction then the basis of choice would disappear in a few generations. But there is a deeper and more compelling reason. There is no intrinsic advantage to an organism in gaudy feathers or large sexual ornaments. If the choice was made in this way it would not be a good choice; the females would not be choosing genes with survival advantage.

But in the above example what would happen if BB is more attractive than both BB and BB. If the heterozygote is more attractive than both homozygotes the gene B will be at an advantage when rare and will expand in number but then will be neutral when common.

The idea that heterozygote advantage could play a part in sexual attraction is rather more interesting. There are survival advantages in having two slightly different genes rather than identical genes at a given locus. We can understand this most clearly in relation to infection (see chapter 3). For instance regulatory sequences of genes control the strength of response in infection. Sometimes it is advantageous to have a strong response and sometimes it is better to have a less strong response. The individual who can be strong or less strong (heterozygote) is at an advantage compared with only strong or only less strong (homozygotes).

There is some evidence that polymorphisms in the HLA system have a role in sexual attraction. This is a form of heterozygote advantage. The HLA molecules are important in influencing the way we respond to infection. When we encounter bacteria they are engulfed by macrophages which take the bacteria into their cytoplasm (phagocytosis) and smash them into pieces (chemically). The proteins that make up the bacteria are cut into fragments approximately 8 to 12 amino acids long (polypetides) and are then posted on the surface of the macrophage in a groove in a HLA molecule. Lymphocytes then inspect the polypeptide chains. A decision is then made:

- This polypeptide fragment is similar to self polypeptides and will not be attacked.
- This polypeptide is foreign but the bacteria are harmless and will not be attacked.
- This polypeptide is foreign and the bacteria are potentially harmful and a clone of lymphocytes will be committed to attacking all bacteria that carry it.

The recognition process involves inspecting the polypeptide within a groove in the HLA molecule and this means that parts of the polypeptide can be obscured by the precise shape of the HLA molecule. The result is that the recognition of specific polypetides is impaired by specific HLA molecules and this can lead to errors in the above decisions. Individuals who have different HLA molecules are at an advantage because they can inspect the polypeptides in different HLA molecules reducing the chance of error.

There is therefore an advantage if we choose a mate with different HLA antigens than ourselves and there is some evidence that this does occur. HLA antigens influence the odours in sweat. We find the odours of members of the opposite sex who are different in HLA antigen type more attractive than those who have similar HLA type.

Non Genetic Aspects

Sexual attraction and mate choice are driven by deep biological forces which we do not understand and over which we have little control. Those forces, generating love and lust, are part of our biological constitution and are ultimately specified by our genes. But to what extent do culture, upbringing, experience, religion and other social forces shape our views and influence our choice.

- Fashion

The phenomenon of fashion indicates that it is not only genes that influence our concept of what is attractive. Last year's hair style, last year's clothes, even last year's opinions enhanced physical attraction last year but not this year. How can something so deeply significant as physical attraction be so ephemeral? I am afraid I cannot answer the question but I would say that I doubt that fashion is as important as we are sometimes led to believe.

Attractive people change their hair style and change their clothes with fashion but remain attractive. Last year's attractive people are this year's attractive people but in different clothes. In fact it is attractive people who tend to lead the fashion, changing because we like change. Marie Antoinette is reputed to have said "there is nothing new only that which has been forgotten". She was not referring to politics, philosophy or religion but to hats. Fashion goes round in circles, it is superficial; it does not influence our underlying concepts of physical attraction.

In the past the poor were thin and only the rich had access to enough food to become fat. This has led some to claim that in the past fatness was beautiful whereas now thin is beautiful. I doubt it. The fact that fat potentates had large harems does not mean that fatness was physically attractive although it might have been economically attractive. Henry the eighth was tall and athletic as a youth but became obese in middle age. He did not become more attractive to women as he got older, quite the reverse. The athletic frame, with broad shoulders and narrow waist, has been attractive in men throughout history. In women the ideal is curvaceous but narrow waist and not fat. The anorexic, stick like models on the cat walk are a perversion.

- Economics

Females are primarily responsible for nurturing the young but males supply 50% of the DNA and variable economic assistance. In the case of the lek, males supply DNA but then have no role in rearing the young. In this situation the females should go for the best possible genes and this means a polygamous pattern of mating. But in monogamous unions males supply not just DNA but also considerable economic assistance. In birds that form monogamous unions males help in building nests and obtaining food. In this situation the females have to compromise, they cannot all mate with the most attractive males. In practice the most attractive males mate with the most attractive females and those lower in the attractiveness stakes have to compromise.

Studies of birds in the wild which form monogamous pairs indicate that mating patterns can be complicated. Tests on the progeny of a given female indicate that not all are sired by the same male partner. A minority of the young are the product of extra-pair copulations. In some studies it has been noted that the progeny of illicit unions are healthier and more successful than the progeny of the monogamous pair. It appears that the female pairs with an average male but seeks "superior genes" elsewhere. In this context I suspect that superior genes mean "fewer deleterious mutations".

- Altruism

One of the problems in evolutionary theory is "how do genes for altruism arise and survive?" If one member of a species loses its gene for altruism it will be at an advantage compared with other members of the species. Genes for altruism should disappear.

But there is no such thing as a gene for altruism, there is only a complicated pattern of behaviour controlled by very many genes that is altruistic. The same genes used in different combinations control other aspects of behaviour which are essential to survival.

Females that form monogamous unions will seek males who will co-operate in the process of rearing young and who will be protective to the female and to the young. This is a form of altruistic behaviour and in evolutionary terms it is advantageous to the male. Is the female hard wired to recognize altruistic behaviour or does she learn through experience? The answer as usual is probably both.

The particular difficulty that evolutionary theorists have is the idea of males showing altruistic behaviour to unrelated males and females, thereby helping them to pass on their genes and in a world of finite resources putting themselves at a disadvantage. But if females are to recognize altruism in males prior to choice then the males will have to be altruistic to others not just the female partner. Thus females will be attracted to altruistic behaviour which is generalized, although the female will expect it to be focused on her and her young once choice has been made.

- Personality

There is no doubt that personality matters when humans choose their partner. We need to get along for a very long time. Whether we choose like or unlike I will leave to others to debate, but that we are influenced in our choice by personality is without doubt. Furthermore I do not know if the choice is hard wired or the product of experience, probably both.

Conclusion

The thesis advanced in this chapter is that genetic factors are important in sexual attraction and in mate choice. Males and females can recognize members of the opposite sex who have low levels of deleterious mutations in the genome and are attracted to them. This basic biological force operates in all animals and is essential in preserving the genome against degradation by mutation.

Neutral mutations have a lesser role, but there is heterozygous advantage in relation to infection and sexual attraction operates to increase HLA polymorphism.

Many other factors apart from genetics influence sexual attraction and some of them are noted above. Social, economic and cultural issues are important as are aspects of personality.

Chapter Seven

Overview

Introduction

The thesis in this book is that information is to biology what energy is to physics. We can analyze biological behaviour using concepts from information theory and make predictions and generate questions for scientific study.

The laws of information theory are:

- Finite capacity: all information processing systems have a finite capacity.
- Uncertainty: information is processed in noise and there is a finite possibility of error.
- Entropy: systems decay with time according to the laws of entropy and therefore the error rate will rise.
- Redundancy: complex systems need a high level of redundancy in order to reduce error rate.

A highly redundant complex system decaying at random with time according to the laws of entropy will show an error rate that is low initially but then rises rapidly in later life prior to final failure. This is similar to the prevalence of degenerative disease and the risk of death in man. The risk of ischaemic heart disease is low in early life but then becomes rapidly more common in later life. The mortality rate shows a similar pattern (Fig 3.2). The way we grow old, acquire disease and eventually die is an inevitable consequence of the laws of information. We could – in theory – slow the process by recognizing and avoiding factors in the environment that lead to decay but we will never abolish the process.

The chance of error on first exposure to a common microorganism, however, gives rise to a different age incidence curve. First encounter with a common organism is likely to occur early in life and the chance of first exposure therefore falls with time (Fig 3.1). But the chance of making a mistake on first exposure rises with time. The result is an age incidence curve of disease that rises then falls (Fig 3.3). If the organism is very common the age incidence curve of disease has a peak which is early and low. If the organism is less common the peak is later and higher. This explains why diseases such as glandular fever, which are caused by common viruses, are rare in childhood, are common in adolescence and do not occur in later life. It also explains why later exposure to microorganisms because of improved hygiene can lead to an increased risk of disease.

One of the advantages of information theory is its emphasis on interaction between the environment (supplying information) and the system that processes the information. The environment also leads to decay in the system; but conversely the networks, specified by genes, can learn from the environment during development to modify their structures in order to optimize performance. The last point is best illustrated by the way that neural networks learn from experience. Thus we avoid the simple mistake of thinking that disease is caused by our genes or by the environment. It is always an interaction between the two. Equally our personality, intelligence, behaviour and health are products of the genes we inherit and the environment in which we develop and live.

Neutral and Deleterious Mutations

A key distinction to make is between neutral and deleterious mutations. This is seen most clearly in discussion of the role of genetic factors in intelligence. There is good evidence that genetic constitution influences scores on IQ tests. There is also the disquieting observation that American Blacks do worse on IQ tests that American Whites. Those who jump to the conclusion that Blacks are genetically inferior to Whites are wrong, but so too are those who conclude that IQ tests are of no value. The correct answer lies elsewhere.

Scores on IQ tests are normally distributed. The range of scores for Blacks, Whites and Asians is the same but Blacks have the lowest mean, that for Whites is intermediate and Asians have the highest mean score.

Neutral mutations arise in spermatogenesis or oogenesis. They occur by chance and are passed onto the next generation by chance. They do not confer any selective advantage or disadvantage; instead their fate depends on chance alone. If they survive and expand within the population then they become part of the genetic constitution of the group in which they arose. If groups are separated for many generations and do not intermarry then they will acquire different neutral mutations and therefore the genetic difference between races depends on neutral mutations. Neutral changes in genes or regulatory components of genes are not better or worse they are equal but different. Thus it is unlikely that neutral changes could lead to variation in the overall score on an IQ test. If American Blacks do worse on IQ tests than American Whites it is not due to the neutral mutations that differentiate Blacks from Whites.

The genetic basis of IQ test scores is much more likely a consequence of deleterious mutations. If a complex network is specified by genes then removing some genes at random will not improve performance; on the contrary removing genes will impair performance. In every racial group there is a balance between the new mutation rate per generation (P) and the mean number of mutations in the genome (Q). Increase in P leads to an increase in Q; a fall in P leads to a fall in Q but it takes several generations before a new equilibrium point is reached. There is no reason to believe that P is constant worldwide and therefore different racial groups might well have different values of Q; if so it is likely that the mean IQ score will vary for this reason.

If the difference in IQ scores is genetically determined then it will be deleterious mutations not neutral mutations that are involved. Deleterious mutations are due to the environment (social, geographical, economic etc) that the group has endured over several generations. If the environment is improved and P falls then Q will also fall and any differences in IQ will disappear. Deleterious mutations are not a permanent part of genetic constitution.

A similar problem arises in relation to health inequality within a population. There is a gradient of social and economic advantage in all societies and invariably it correlates with health. Individuals who enjoy advantage in social and economic circumstances also enjoy the best health and live the longest. Those at the bottom of the social scale have the worst health and the shortest lives. Any suggestion that genetic factors could be involved is quickly rejected by some, for the best of motives, because the idea that the poor are genetically inferior is not acceptable. But it would be odd if genetic factors were not involved because they are in every other aspect of health and disease.

The answer once again lies in differentiating between neutral and deleterious mutations. The poor are not genetically inferior in any society but they are likely to have more deleterious mutations. (If that seems a contradiction remember that the vast majority of mutations in the genome are neutral; furthermore deleterious mutations are but temporary residents in a breeding population). Deleterious mutations impair health, intelligence and physical attractiveness. The number of deleterious mutations in the genome shows a Poisson distribution (approximate) within a breeding population. It is inevitable that those with more deleterious mutations will gravitate towards the socially and economically deprived end of the spectrum.

Those who seek to deny the role of genetics in health inequality do it for the best of motives. They see genetic influences as permanent and in effect an excuse for tolerating gross health inequality; whereas they wish to do the opposite – reduce inequality of all forms. I contend that the best way of reducing inequality is to work out which aspects of the environment cause mutations to occur. If we could reduce P then Q would fall over a number of generations and the entire population would be healthier, more intelligent and better looking. This would not abolish inequality but the gradient would be less steep.

Intelligence, Health and Physical Attraction

Deleterious mutations in the genome interact synergistically to impair every aspect of biological behaviour and performance. They will impair the development of a symmetrical body form, performance on tests of ability and the action and success of systems that preserve health and prevent disease. In other words health, intelligence and physical attraction will be closely correlated or at least that is the prediction of the ideas developed in this book.

There is evidence that health and intelligence are closely related. Those who do well at school advance socially and economically and enjoy better health than the average. But there is less evidence to support the idea that physical attraction is linked to health and intelligence.

The difficulty with studying the links between these three is there are many confounding factors. If we know somebody is intelligent we tend to judge them as more attractive. If we find somebody attractive we often judge them to be intelligent. Disease can directly impair attractiveness. The age of the observer and the observed influences the perception of attractiveness. Social factors such as accent, style of dress and deportment will influence our assessment. Thus designing a study to assess physical attractiveness and relate it to intelligence is not easy, but it can be done.

Beauty Is Not Just Skin Deep

One of the intriguing ideas to emerge from the analysis in this book is the central role that beauty plays in biology. The argument is as follows:

- Mutations are inevitable and those that affect coding regions are most commonly deleterious.
- The only way to reduce the number of deleterious mutations in the genome is by sexual reproduction. In asexual organisms deleterious mutations accumulate but some of the progeny of a sexual union have fewer deleterious mutations than the parents.
- There is biological advantage to a species if one gender can recognize and be attracted to members of the opposite sex who have low levels of deleterious mutations.
- Recognition of low levels of deleterious mutations could be based on utility or aesthetics.
- Survival depends on utility and therefore biology needs complex redundant systems to preserve utility. That means one cannot detect low levels of deleterious mutations by assessing utility.
- Aesthetics has no intrinsic survival value therefore it can be used to signal low levels of deleterious mutations.
- But the system only has biological advantage if the genes that specifiy networks for utility also combine differently to specify networks for beauty.

Thus walking, running and climbing are useful attributes but it is only the male that walks, runs and climbs with grace, style and panache that will excite the attention of females. Equally bright plumage is attractive and though of no intrinsic utility it indicates functioning genes that will increase the chance of healthy offspring. Thus in nature the competent survive but only the beautiful pass on their genes to the next generation.

A beautiful object, or pattern or movement is one with a high level of information and a low level of uncertainty. If the pattern is copied without error it remains beautiful. If random errors are introduced then the pattern becomes less attractive; it also loses information and gains uncertainty. This must be a universal rule otherwise sexual attraction throughout nature would not work as suggested. It works because aesthetics indicates a high information system with a low error rate. If it is a universal system then we share our concept of beauty with the rest of nature. Perhaps that is why nature is beautiful to our eyes; because in nature only the beautiful survive.

The biological basis of beauty is rooted in physical and sexual attraction for members of the opposite sex but its existence leads to a much broader concept. The same mechanism leads to attraction between members of the opposite sex for all animals, but we are only sexually and physically attracted to members of our own species. Thus sexual attraction and beauty can be separated; the latter is a broader concept the former very specific. But if beautiful things are attractive then we obviously will want to surround ourselves by objects of beauty. We want beautiful buildings, beautiful gardens and beautiful pictures. We listen to music, read poetry, prefer elegant prose to the utilitarian and like to dance. None of this is remotely useful.

In fact an objective view of our behaviour would indicate that we live in order to enjoy aesthetic pleasures; because biology has designed us in that way.

Microbial Flora

In the first year of life we encounter a vast range of bacteria, many of which become permanent members of our microbial flora. There are between 400 and 800 different species of bacteria that can be found in the gastrointestinal tract and one individual has between 40 and 80 at any one time. The proteomic complexity of the bacterial flora is as great as our own proteomic complexity. The task of the immune system is to distinguish self from not self proteins and to differentiate between pathogens and commensals. Three types of mistake can occur:

- Failure to recognize and respond to a foreign protein on a pathogen increases the risk of infection (an error of omission).
- Responding to a protein which is similar to self risks autoimmune disease (an error of commission).
- Responding to foreign proteins on commensals produces unnecessary inflammation (error of commission).

Information theory indicates that mistakes will occur, they cannot be entirely eliminated, but they can be reduced. In theory, low dose, early, mucosal exposure should lead to a reduction in errors.

Low dose exposure gives the immune system time to analyze bacteria and make the crucial distinctions between pathogen and commensal, and between self and not-self proteins. The information processing system has a finite capacity and therefore more time means fewer errors.

Information processing systems deteriorate with time and therefore early exposure is the time of lowest risk.

Most of the information processing capacity is on the mucosal surfaces, particularly of the gastrointestinal tract – more capacity means a lower error rate. Furthermore once the gastrointestinal flora is established the growth of new pathogens is inhibited and this gives the immune system longer to analyze the bacteria, again reducing errors.

This poses a major challenge to scientific research. We need to map the microbial flora in detail and establish how it changes with time. We need to discover which proteins on which bacteria are stimulating clones of lymphocytes that cause autoimmune and allergic disease. We must then determine the optimum time, dose and route of administration of those bacteria so that the risk of error is reduced to a minimum. Ultimately this means manipulating exposure to bacteria in the first few years of life.

Study of the human genome is gradually revealing what we could have deduced from first principles: the neutral mutations that are the main cause of genetic polymorphism are linked to infection. Neutral mutations render us more susceptible to some organisms but less susceptible to others – they are neutral overall. But this leads to heterozygous advantage – with genes linked to infection it is often better to have two different genes at a locus than two the same. Females have more heterozygous advantage than males because they have two X chromosomes and males have only one. The X chromosome carries 5% of the genome giving females a considerable advantage in fighting infection. But there is a twist in that if one uses two strategies rather than one to deal with bacteria errors of omission will be reduced but errors of commission could be increased – could that explain why females are less susceptible to infection but more at risk of autoimmune disease?

In the Western World the risk of death in females is less than that in males for every year of life from infancy to old age. Women die of ischaemic heart disease and strokes as do men, but in women death comes later. In view of women's advantage of two X chromosomes it is tempting to speculate that infection could have a role in damaging endothelial cells and predisposing to vascular disease. Time will tell; but once again the information approach generates questions for scientific study.

I will hazard a prediction: a major area of scientific endeavour over the next 50 years will be gaining a deeper understanding of the microbial flora, learning to control its acquisition and modify its composition. This will produce benefits not only in infection and immune mediated disease but also in many other conditions which we currently do not regard as the consequence of infection. The next half century will see the rediscovery of the germ theory of disease.

Monitoring Mutation Rates

Complex behaviour such as eating, sleeping, talking, thinking and dreaming is controlled by vast networks of genes. Genes specify proteins but there are also complex regulatory elements which put genes together in different combinations to do different tasks. This allows enormous complexity with only 25,000 genes; they can be combined in limitless ways to perform a vast range of actions. Systems of this complexity need a high level of redundancy.

When cells divide at mitosis there is a low but finite rate of mutation which cannot be eliminated. Exposure to chemical mutagens and radiation will increase the rate of mutation as will a lack of anti-oxidants in the diet. A good example of a mutagen is tobacco smoke. Smoking increases mutation and avoiding cigarette smoke will reduce it.

The majority of new mutations that affect protein coding elements or regulatory elements will be deleterious (but the majority that survive long term in the genome will be neutral). They lead to functional loss of a gene, or part of a gene or a disturbance of regulation. In a highly redundant system the loss of one or two elements will have no measurable deleterious effect – the essence of redundancy. But loss of further elements will start to have a measurable effect and crucially there is synergistic interaction between the deleterious mutations so that just one or two more mutations in the system will lead to breakdown. This is a general property of a complex highly redundant information processing system.

Mutations occur during mitotic divisions which lead to the production of spermataozoa and oocytes. The best estimate is that in the Western World at the current epoch parents pass on average one new deleterious mutation to their offspring in each generation. Without selection that would lead to a steady build up of deleterious mutations in the genome and eventual extinction. We are saved from extinction by the process of synergistic interaction but there is a price to pay. If P is the new mutation rate per generation and Q is the mean number of mutations in the germ line of individuals then the best estimates are P=1 and Q=8. The precise numbers, however, do not matter but the principle does. During meiosis deleterious mutations are distributed at random to oocytes and spermatozoa and then these gametes fuse at random to form zygotes. The zygotes in turn develop into embryos and then feti and finally children and adults. The result of random distribution of deleterious mutations to zygotes is that the mean number of deleterious mutations will be 9 (parents have mean of 8 and 1 additional arising in gametogenesis i.e. P+Q)   but the actual number of mutations in any one zygote will vary on either side of 9, some as low as 0 or 1, a few as many as 18 or 19. This distribution is given the name of a Poisson distribution.

The zygotes with fewer than 8 deleterious mutations are likely to survive and develop into adults; but those with many more than 8 will suffer the effects of synergistic interaction and will fail to develop. Thus the survivors have a mean of 8 and the population is in balance but the range of mutations will be from 0 or 1 to maybe 10 or 12. Those who have few deleterious mutations will have genetic networks which function at a high level; those with more deleterious mutations will experience some impairment of function. The result is a gradient of intelligence, health and physical attractiveness.

If in a population the mutation rate falls due to avoidance of chemical carcinogens or radiation or to improvement in diet then the mean value of Q will gradually fall over several generations. The result will be an improvement in fertility as fewer zygotes fail to develop, and those that do develop will be more intelligent, healthier and better looking than their parents. Furthermore the gradient of inequality in the population will become less steep, although it will never disappear.

Another challenge for medical science over the next half century is to develop methods to monitor mutation rates in human populations and seek to recognize and avoid those factors which increase the mutation rate. Over the last 250 years in the developed world there has been a gradual improvement in the health of the population, average height has increased, childhood mortality has fallen, average lifespan has increased and over the last 50 years measured IQ has risen. There are many reasons for these changes but a fall in mutation rate due to less environmental pollution and improved year round diet could have played a part. But we cannot be complacent because industrial societies produce a wide range of chemicals which could reverse the process. We must learn to measure, monitor and intervene so that our children are healthier, more intelligent and better looking.

Ageing

We are healthy when young but as we grow old we become increasingly likely to become ill, develop disability and die. This is the ageing process and it is inevitable although the rate at which deterioration occurs is not fixed.

Ageing is a consequence of the third law of information – that systems decay at random with time and the error rate of information processing will rise. But the rate of decay is modified by redundancy producing the classical curve relating risk of death to age (Fig 3.2). Environmental factors which increase the rate of somatic mutation will cause information loss and will speed up the process of ageing. Equally avoiding mutagens, such as tobacco smoke, will retard the process.

It is interesting to note that exposure to cigarette smoke increases the risk of most diseases. The simplest way to explain this is that smoking increases somatic mutation in our cells. It speeds up the rate at which we grow old so that ischaemic heart disease develops early and degenerative lung disease is accelerated etc. In addition the collagen support to the skin is damaged early and wrinkles develop faster and are more pronounced.

If somatic mutation is a key part of ageing then it provides another reason for learning to measure and monitor mutation rates. We cannot prevent ageing but less mutation would slow the process down. We would live longer and healthier lives. There is a public health imperative to know the rate at which our cells are being damaged by the environment in which we live.

Future Research Strategy

- Measure and monitor the rate of somatic mutation in the general population.
- Control microbial exposure to ensure it is low dose, early and mucosal.
- We are biologically programmed to seek aesthetic pleasure and attempts to maximize happiness must be based on this principle; happiness will not come from economic advance alone. We prefer beauty to utility.

Appendix

The book is based on a series of articles written by the author over many years. There are no references in the text but those who wish to delve more deeply into the subject matter will find the articles, listed below, a good place to start.

Morris JA. Autoimmunity: a decision theory model. Journal of Clinical Pathology 1987; 40: 210 – 215.

Morris JA, Haran D, Smith A. Hypothesis: common bacterial toxins are a possible cause of the sudden infant death syndrome. Medical Hypotheses; 1987; 22: 211 – 222.

Morris JA. A mutational theory of leukaemogenesis. Journal of Clinical Pathology; 1989: 42; 337 – 340.

Morris JA. A possible role for bacteria in the pathogenesis of insulin dependent diabetes mellitus. Medical Hypotheses 1989; 29; 231 – 235.

Morris JA. Spontaneous mutation rate in retinoblastoma. Journal of Clinical Pathology 1990; 43: 496 – 498.

Morris JA. The age incidence of multiple sclerosis: a decision theory model. Medical Hypotheses 1990; 32; 129 – 135.

Morris JA. A mutational model of carcinogenesis. Anticancer Research 1991; 11: 1731 – 1734.

Morris JA. Low dose radiation and childhood cancer. Journal of Clinical Pathology 1992; 45: 378 – 381.

Morris JA, Ageing, information and the magical number seven. Medical Hypotheses. 1992; 39: 291 – 294.

Morris JA. Childhood cancer around nuclear installations. European Journal of Cancer Prevention 1994; 3: 15 – 21.

Morris JA. Information and clonal concepts in ageing. Medical Hypotheses 1994; 42: 89 – 94.

Morris JA. Information and observer disagreement in histopathology. Histopathology 1994; 25: 123 – 128.

Morris JA. Psychological influences in disease; a decision theory model. Medical Hypotheses 1994; 43: 319 – 321.

Morris JA. Stem cell kinetics and intestinal cancer. Anticancer Research 1994; 14: 2073 – 2076.

Morris JA. The scandal of poor medical research: theory must drive experiment. British Medical Journal 1994; 308: 592.

Morris JA. Schizophrenia, bacterial toxins and genetics of redundancy. Medical Hypotheses 1996; 46: 362 – 366.

Morris JA. Genetic control of redundant systems. Medical Hypotheses 1997; 49: 159 – 164.

Morris JA. Fetal origin of maturity-onset diabetes mellitus: genetic or environmental cause? Medical Hypotheses 1998; 51: 285 – 288.

Morris JA. Information and redundancy: key concepts in understanding the genetic control of health and intelligence. Medical Hypotheses 1999; 53; 118 – 123.

Morris JA. The common bacterial toxin hypothesis of sudden infant death syndrome. FEMS Immunology and Medical Microbiology 1999; 25: 11 – 17.

Morris JA. Effects of somatic cloning. Lancet 1999; 354: 255.

Morris JA. The kinetics of epithelial cell generation: its relevance to cancer and ageing. Journal of Theoretical Biology 1999; 199: 87 – 95.

Morris JA. How many deleterious mutations are there in the human genome? Medical Hypotheses 2001; 56: 646 – 652.

Morris JA. Information theory: a guide in the investigation of disease. Journal of Biosciences 2001; 26: 15 – 23.

Morris JA. Information: theoretical and applied. Pages 165 – 183. In Progress in Pathology, volume 5, edited by Kirkham N, Lemoine NZ. Greenwich Medical Media Ltd, London, 2001.

Rigby JE, Morris JA, Lavelle J, Stewart M, Gatrell AC. Can physical trauma cause breast cancer? European Journal of Cancer Prevention 2002; 11: 307 – 311.

Morris RD, Morris KL, Morris JA. The mathematical basis of sexual attraction. Medical Hypotheses 2002; 59: 475 – 481.

Deolekar MV, Morris JA. How accurate are subjective judgments of a continuum? Histopathology 2003; 42: 227 – 232.

Morris JA, Morris RD. The conservation of redundancy in genetic systems: effects of sexual and asexual reproduction. Journal of Biosciences 2003; 28: 671 – 681.

Morris JA. Common bacterial toxins and physiological vulnerability to sudden infant death: the role of deleterious mutations. FEMS Immunology and Medical Microbiology 2004; 42: 42 – 47.

Morris RD, Morris JA. Sexual selection, redundancy and survival of the most beautiful. Journal of Biosciences 2004; 29: 359 – 366.

Morris JA. Synergistic interaction of heterozygous deletions impairs performance and confers susceptibility to disease at all ages. Medical Hypotheses 2005; 65: 483 – 493.

Morris JA, Harrison LM, Biswas J, Telford DR. Transient bacteraemia: a possible cause of life threatening events. Medical Hypotheses 2007; 69: 1032 – 1039.

Morris JA, Harrison LM. Hypothesis: increased male mortality caused by infection is due to decrease in heterozygous loci as a result of a single X chromosome. Medical Hypotheses 2009; 72: 322 – 324.

Morris JA, Harrison LM, Brodison A, Lauder R. Sudden infant death syndrome and cardiac arrhythmias. Future Cardiology 2009; 5: 201 – 207.

www.ingramcontent.com/pod-product-compliance
Lightning Source LLC
Chambersburg PA
CBHW060035210326
41520CB00009B/1139